Are We Just Bubbles?
An Alternate View of Existence

Are We Just Bubbles?
An Alternate View of Existence

Daniel P. Bowlds

Cover Art & Cartoons: Kristen Terrana-Hollis

Daniel P. Bowlds
2015

First Printing: 2015

ISBN 978-1-329-59041-0

Daniel P. Bowlds
Lewisport, KY 42351

This book is dedicated to all people of the Earth who seek the truth of our existence with humility; be they rich or poor, young or old, male or female, of any race, nationality, religion, political persuasion, or of any profession from the creative artists, ministers, scientists, engineers, teachers, doctors, ... to the men and women laborers working to feed and shelter themselves and their families.

Contents

Acknowledgments

First of all, I would like to thank my parents for bringing me into this world and for providing a good environment for me while I was growing up. Although we weren't financially wealthy, I always had enough to eat, a warm place to sleep, and an abundance of love and support from my family. And even though my dad wasn't particularly interested in science when I was young, a statement he made to me while star gazing one night, about the stars and Universe going on forever, has stuck with me and has been a lifelong source of awe and fascination.

And, I owe an enormous amount to my loving wife Elaine, who from the very start of our relationship has had great faith and trust in me, who has stood by me through the thick and thin of raising up our six children while I was away working and going to school, who has cooked and cleaned, washed dirty diapers, wiped dirty faces, kissed their "owwees", and has tucked them in bed. She has done this with love and without complaint, and has been there as a "sounding board" for all of my crazy ideas and ramblings. Truly, I would not be writing this book if it wasn't for her love and support.

And, I am grateful for my employer General Electric Company who provided tuition support, and an opportunity to attend classes during part of the working day so that I could finish up my degrees. And I'm thankful for the support and mentor-ship received from the engineers at my work. I'm thankful for Brescia University, and for the Mathematics, Engineering, Physics, Religion, History, and Creative Writing professors who taught me.

And, I also owe a tremendous amount of thanks to my Kindred Spirit friends who have been a great source of inspiration and support for me in the last ten years; the dreamers, creative and imaginative thinkers, the artists, singers, songwriters, poets, dancers and actors. These have a large compassionate heart with a great capacity to love. They have helped me to see that we are all connected, and that we must make a giant leap into the next level of our understanding, consciousness, and social evolution if we are to survive on this planet. Without their kindness, love, willingness to seek truth, and tolerance of others, I would never have come to this realization.

*

"Willful Ignorance and Arrogance are Ego's evil twins. Where one is found, the other is likely to be. We cannot already know that which we do not know. And since we can only probe the depths of our ignorance with acquired knowledge, it will be the humble who wisely use knowledge for greater understanding. Those who arrogantly close their minds in undoubted certainty of their own knowledge, doom themselves to live in ignorance!"

Daniel P Bowlds

"A man may imagine things that are false but he can only understand things that are true for if the things be false, the apprehension of them is not understanding."

Isaac Newton[1]

"I most seriously believe that one does people the best service by giving them some elevating work to do and thus indirectly elevating them. This applies most of all to the great artist, but also in a lesser degree to the scientist. To be sure, it is not the fruits of scientific research that elevate a man and enrich his nature, but the urge to understand , the intellectual work, creative or receptive."

Albert Einstein[2]

[1] Isaac Newton from his Theological Notebook (Part 1) Source: Keynes Ms. 2, Kings College, Cambridge, UK published online Sept. 2003 by The Newton Project

[2] Albert Einstein from his book "The World as I See it", an authorized translation of "Mein Weltbild" distributed by Citadel Press

Preface

This little bitty book about the great big Universe presents a different concept of our physical reality. We made some basic assumptions about measurements of length and time a very long time ago, and have never questioned them since. It is possible that these assumptions might not be true, and since we have based all of our science and understanding about physical behavior upon these measurements, our concepts of how the Universe works may also be in error. My intent in this short writing is to humbly present the possibility of another way in which the Universe could exist and still behave as we observe it with our measurements. Some of you with strong convictions about the certainty of your own knowledge are likely to immediately reject these ideas, but if you have an open mind and are willing to consider other concepts, then you may gain a deeper insight into the reality of what we are. Having said that, I do not presume to be any kind of authority. I will be humbly presenting the possibility of an unseen dimension and expansion of space that might be the underlying cause of all that the Universe is. It is not my intent to "prove" any of what I am proposing. But if what I propose cannot be disproved, then the certainty of what we now think the Universe is, based upon our measurements, comes into question.

I will briefly mention other things here in the preface that do exist but which I believe are not contained by or solely within the physical Universe. I think it is necessary to distinguish these things here so that we don't confuse them while trying to understand the mechanics of the physical Universe.

First of all, I believe that in order to be part of or contained within the bounds of the physical Universe, those contents have to exhibit physical properties and interact with other known articles contained within it. Nonphysical things that exist such as logic, reason, knowledge, mathematics, software, etc. have no affect on the physical Universe (note that they exist primarily in the conscious minds of living things). The properties of matter and its interactions (gravity, time, mass, or energy) do not affect the nonphysical things (i.e. gravity does not influence the fact that two plus two equals four). Therefore the nonphysical things are not strictly bound by the Universe. This is not to say that the nonphysical things cannot exist in the physical Universe, or in the case of life, do not use the things in the physical Universe as hosts for its continued presence in it, it just means that these things can exist outside the bounds of the Universe, and are not changed by the physical laws of the Universe while they are within it.

Now this area of philosophy is typically where Religion and Science part, and has been the cause of much political strife. From the strictly scientifically minded perspective, everything that exists comes from and is caused by the physical properties and contents of the Universe. It is assumed that the physical laws of behavior and the properties of matter in the Universe are constant and unchanging. Thought, will, and opinions are just a byproduct of life (which is assumed to have come from the Universe). Both the Universe and Life just "happened" by some fluke of statistics or chaos (even though these nonphysical things have no affect on the Universe and this conclusion might be a deviation from the scientific method).

From the religious perspective, everything in the physical Universe came from and is caused by some ever existing, all knowing, Almighty Power (God). Things in our conscious reality that exist, but are not physical, are called things of the spirit. These things are like God and were created before the physical Universe was created. Since the cause of everything is God, God can alter the behavior of the Universe at will, suspending the natural laws that was created. There is also a sense that the angel and devil spirits, and even the human spirit (since they are like God) can have some power over the physical reality.

As our consciousness took shape out of the primordial fog, we confused the two realities and lumped everything together. Mythological stories might have said, "The Evil Spirit came upon the World (their Universe) and the Sun was smitten". In actuality, a volcano might have erupted five hundred miles away and produced an ash cloud that blocked a portion of the sun light. Our reasoning associated human traits with the spiritual God and other spirits in order to explain the physical actions we were seeing. When bad things happened to us physically, we thought that God did this directly as a punishment for something we had done, or was allowing some lesser evil spirit to control the physical circumstances we were in because we were bad. We thought God must be behaving like some kind of Almighty Tyrant. We gave God human traits such as anger and vengeance to explain why these things happened to us, or was allowed to happen.

As our thinking became more distinct, and we were more observant (probably because we had more leisure time), we began to see that some things in the physical reality behave consistently in certain ways, regardless of what we think or do. The study of this behavior became Science, and the "Religion" of Science came to be that we only acknowledge the existence of things that exhibit consistent physical behavior. The scientific method developed in order to keep opinions from influencing the interpretation of the physical behavior, and strict rules developed so that judgments were based upon direct observation and actions that had repeatable results. A "Cardinal Sin" of science is to selectively present data from tests to confirm an opinion while callously ignoring other data in the tests that might contradict the opinion (This was not necessarily the case for the religious mythological record though, which was shaped to present an ideology). And so it became easy for those engrossed in science to forget and deny that there might be a Creation that is something other than from the physical Universe, or even that the Universe had a Creation (a beginning).

And so the rift has widened between religious and scientific thinking. It has now come to be defined by political forces on both sides (those who wish to dominate and control the actions and thinking of others), such that religion and science are becoming mutually exclusive disciplines. We have religious leaders who insist that God created the Universe in seven earth days and they ignore all physical evidence to the contrary, and we have atheist scientists who think that the Universe and life in it just "happened". The mathematical singularity at the beginning of the Universe is explained away by saying we know what happened right after it was created. And so it must have come from a dimensionless point with infinite mass (came from nothing) because we are so "close" to seeing the beginning of it.

If we can step back from this for a moment, I think we can see both the physical and spiritual realities. I believe that both exist and are separate creations. They do not interact with each other, even though they are dependent upon each other. Does the Universe exist if there is nothing conscious of its existence? Can consciousness, intelligence, and reasoning exist in the Universe without being confined as life into some physical medium? As long as we are human, both physical (that which is bound to the physical Universe), and spiritual (that which is not bound), we may never be able to solve these mysteries. For now though, I am going to talk about the physical Universe that we know and live in, and the things that exhibit physical properties within it. We will use the non-physical "spiritual" parts of our creation such as reasoning, logic, and mathematics to describe the actions we are seeing, and try not to influence the observances with opinions.

Chapter 1: Measurements & Units of Measure

All that we know in our conscious minds about our surroundings has primarily come to us through our senses. From birth, we have touched and felt, heard, smelled and tasted, and have seen all sorts of objects and things to form opinions of. We learn the concept of distance, how heavy things are, learn about hot and cold things, about loud and quiet things, about light and dark, etc. We have been able to extend our sense of sight artificially with all sorts of instruments; x-ray machines, microscopes, telescopes, atom smashers etc. so that we can learn more about the behavior of the things we can't normally see. With these instruments we have been able to look inside the atom to see what it consists of (somewhat), and we have seen the light from a million galaxies from a distant past time and an unfathomable distance away. We have established means to measure our observations consistently and found that nature seems to follow certain laws as we observe the behavior of things.

The three basic dimensions of measure that scientists presently use for observations are: length, mass, and time. In the MKS system of measurements these dimensions are the Meter, the Kilogram, and the Second. All other expressions used to describe natural phenomena are either unit-less, or combinations of the three basic units. For instance, velocity or speed units equals distance divided by time, force units equals mass times length divided by time squared, and so on. We will now examine the three basic dimensions in the following sections of this chapter to see if we really know what they are and how certain we are about that knowledge.

1.0 Length Measurements

All of us have used a ruler at times to measure the length of something. Maybe you measured your waist size, or a piece of lumber, but have you ever stopped to consider what it is you are actually doing? Since we are using this concept of length to observe our surroundings, let's go back and look at this again.

Finite Length

When measuring the length of something, the very first thing you unconsciously do is to assume that the object you are going to measure is finite in length (has a beginning and end) and that it is constant (not changing in length while you try to measure it). Then you take another finite and constant object which was set up as a measurement standard length (a ruler), and compare that to the length of the object being measured. For instance, to measure a line drawn on a piece of paper, you have to find the line's beginning and ending points. Then to measure the line in some measurement units, you take the fixed length reference (a ruler) and compare it to the line's length. The comparison is done in the form of a ratio, and the reference's name becomes the units of measurement (inches, meters, etc.). These units are then assigned to the ratio.

$$\frac{(Object\ Length)}{(Reference\ Length)}\ Reference\ units$$

When we say that a line is 0.1 cm long, what we really mean is that the line *starting* at zero *ends* at a fractional ratio of 1/10 of our reference's one centimeter length. The one centimeter reference is based upon a tangible unchanging standard. Both the reference and the line being measured have a beginning and an end. Typically, we have a ruler which may be multiple standard lengths long, and this ruler has graduations on it in fractional and integer multiples of the reference length.

Ruler: Finite and Constant Measurement Standard

Figure 1.1

To the left of 1 we have linear fractional divisions to the finest degree, and we define the beginning of the ruler as zero length. To the right of 1 we have linear graduations in integer multiples of the reference and fractional subdivisions between each of them. With the ruler, we can read off the ratio of the measured length to the reference length directly. For instance, to measure a line, the beginning points of the ruler and the line are placed upon one another (coincident) beyond any discernible fractional increment, and we define this as the beginning point (at zero length). The ending point of the line being measured is matched coincident upon the ruler to the finest degree, and the measurement ratio is read directly off of the ruler.

In addition to measuring actual objects that have length, we commonly use the concept of applying a negative length to represent the length of absent objects (subtraction is the addition of a negative amount). For example we might say the board was two inches too long and we need to add a minus two inch board to it (cut it off) to make it the right length. But remember, this is just a way to adjust real objects, and these negative length objects do not actually exist in the physical Universe.

Infinity, vs. a Measurable Quantity

"I'll love you forever and ever and ever!" We've heard this before, and some of us have even said it to the ones we love, but what is forever? It's an idea that time has no end, and yet we have experienced nothing in our surroundings, or even in the Universe that would suggest that time goes on forever. Everything we have seen or experienced has a time when it starts and when it ends, stars are born, live and then die. We think the Universe is "exploding" as time passes and that there was an almost zero time when it began. We don't know if it has an end or not for sure, but we do know that it appears to be expanding and "cooling" as though it will die a frozen death. All physical objects (with mass) have a finite size, and so on. So, let's really look at what infinity means in a finite universe. We will start with the "length" concept of an infinite line.

One thing that may not be so obvious is that an infinite line has no beginning or end. Another thing is that if it exists, it always has a positive length dimension. It is never zero or negative. So how can we represent this infinite line in our finite Universe? The answer is, we will only look at a portion of the infinite line and make a scale on it with measurement units that can expand or contract and "slide" along the infinite line to adjust for whatever fixed finite unit of measurements we are using.

So now, to use this ruler to describe how our Universe might be expanding into an infinite void of non-Universe space, we could pick an arbitrary fixed and constant length standard (a meter for instance), and use that scale on our infinite ruler to measure distances in the infinite space outside of our Universe (hereafter called the Void). Here is what an infinite "ruler" might look like:

Infinity Quantization Ruler for a Finite Length Subset

Figure 1.2

The ruler still has no beginning or end, but we can start our measurement at the finest point discernible on our open origin, and we can extend the other end as far as we need to make the measurement at a distance from our starting point. In doing this however, we have made the assumption that there has always been a fixed and unchanging meter length relative to the Void that our Universe is expanding into. We assume that the beginning point we have chosen has always been there, even though the objects in the Universe we are measuring may not have been there at some point in time.

The assumption that our meter reference is constant and unchanging in absolute space might be false though. Our standards may be unchanging relative to the Universe as we observe them, but that doesn't necessarily mean they are unchanging to the absolute space the Universe is expanding into. This is something that we cannot see and is outside of our realm of observation. More about this later, as this is the little thread that could unravel our whole notion about how the Universe works.

The Ether as a Measurement Medium

The notion of an infinite empty vacuum space disturbed pioneering physicists like Isaac Newton who had a mechanical and deterministic view of how things work in the Universe. How could gravity and inertia work as action at a distance without anything connecting the objects? They really didn't want to think of nonphysical things like "spirits" or an innate nature causing this. So the idea of a stationary background "ether" permeating all of space cropped up as being a plausible explanation. Here is a quote from Newton's letter to Richard Bentley[1] regarding the subject:

> *"That gravity should be innate, inherent, and essential to matter, so that one body may act on another at a distance in a vacuum, without the mediation of anything else, by and through which their action and*

force may be conveyed from one to another, is to me so great an absurdity that I believe no man who has in philosophical matters a competent faculty of thinking can ever fall into it. Gravity must be caused by an agent acting constantly and according to certain laws, but whether this agent be material or immaterial I have left to the consideration of my readers."

Considerable work was done on the ether theory in the nineteenth century to explain the behavior of light as a wave through a medium, and possibly as a cause of gravity and inertia. All of the ether theory work culminated at the end of the nineteenth century with Hendrik Lorentz's and Henri Poincare's work which was done 1892-1895 and finished by Poincare 1899-1904. Shortly after that in 1905, Albert Einstein published his special theory of relativity which did not require an ether to predict the actions that were observed. And so the ether theory was then abandoned by most scientists in favor of a simpler explanation, although the cause of action at a distance was still not known. However, years later Einstein alluded to some kind of motionless invisible medium being possible, and that his theory did not rule out the possibility of its existence. A quote here from his address on May 5, 1920 at the University of Leyden[2]:

"Recapitulating, we may say that according to the general theory of relativity space is endowed with physical qualities; in this sense, therefore, there exists an ether. According to the general theory of relativity space without ether is unthinkable; for in such space there not only would be no propagation of light, but also no possibility of existence for standards of space and time (measuring rods and clocks), nor therefore any space-time intervals in the physical sense. But this ether may not be thought of as endowed with the quality characteristic of ponderable media, as consisting of parts which may be tracked through time. The idea of motion may not be applied to it."

So we can see here that the length dimension, one of the fundamental ways we measure our observations, is referenced to something physical in the Universe, and is considered constant and unchanging in our reference frame. The assumption that our length reference is constant outside the boundaries of the Universe might be false. We cannot know this for sure from here.

2.0 Mass

"The Golden Gate Bridge is massive!" To the nonscientific person, massive invokes an image of something large, dense, and heavy. But what does mass mean in scientific terms? Generally it means the magnitude of the action that inertia and gravity have on a quantity of physical substance that occupies volume in Universe space.

Most people are familiar with the elements which make up the physical objects that we see in our everyday lives. We notice that some of the elements are more dense than others, a block of lead for instance is heavier than a block of aluminum of the same size. Why are they different? The reason is, the lead has many more subatomic particles in its atom than does the aluminum. And since the subatomic particles in both atoms are alike in their inertial and gravitation properties, the elements with a higher atomic number will be affected in direct proportion to it. Even particles like the electron, proton, neutron, etc., occupy volume in the instantaneous sense (this gets fuzzy with time passing due to Heisenburg's uncertainty principle) and have inertial and gravitational properties, although the gravitational force inside the atom is very weak in comparison to the electric and magnetic forces these particles are subjected to. Electrons for instance have a mass of 9.11×10^{-31} kg, exhibit an inertial force when accelerated, and store kinetic energy when moving relative to a stationary object. So a nonscientific person might ask, are the subatomic particles "solid" or are they "fluffy" and composed of even smaller particles? Well, the short answer is maybe. Matter can be much more dense than atoms, that is, particles smaller than an atom can have much higher inertial and gravitational properties than a single atom. This is the case of matter in a Black Hole, and it might be that the subatomic particles making up the atom might also become more dense. So really then, the mass measurement can be summed up in this statement:

The mass of an object is determined by the magnitude of the action that inertia and gravity have on a quantity of physical substance that occupies a particular volume in Universe space.

3.0 Time

"Where does the time go?" Probably everyone has heard this and understood what the person was trying to express. It means for that person, time has passed by quickly and hardly any memories of it remain. But at a deeper level, what is time? Where does it come from? What makes it pass? Would it be passing if we had no memory? Can we stop time or back it up?

This puzzle has plagued philosophers since they came to be. Books have been written about it. Dreamers have imagined what it would be like if we could alter time and move our consciousness back to the past or forward to our future selves. In the last century we have discovered a connection between time and the velocity of matter relative to the ultimate velocity of light, but time is in no way physical that we can see, so how can this happen? We know that all actions, forces, and physical properties have to happen in time to be measured, but we still don't really know what it is exactly or what causes it.

Sometimes scientists and engineers talk about "instantaneous" values of velocity, force, or potential, etc., as if a measurement could be made without time passing. If absolutely *nothing* changed in some way though, would it be possible to tell if time was passing? An assumption was made in talking about instantaneous values that all actions and properties would still exist if time stopped. But, what if everything in our Universe *depended* on time passing to exist? Is our perception of time just a sequential record of changes that was caused by an underlying cause of existence? These are deep thoughts and are likely to rattle us to the bone. So for right now, I will leave this fundamental measurement dimension (the Second) as a period length comparison of a known action in the Universe such as the swing period of a clock pendulum, or the period of an atomic clock vibrating frequency, to the period of some other action taking place. Here again we assume simultaneity and consistency of this dimension throughout the Universe and some fabric of space.

[1] Isaac Newton from his letter to Richard Bently Source: 189.R.4.47, ff. 7-8, Trinity College Library, Cambridge, UK published online Sept. 2007 by The Newton Project

[2] Albert Einstein from his lecture "Ather und Relativitatstheorie" presented at the University of Leyden (Berlin: Springer) on May 5, 1920 (Reference information taken from Aperion, Vol. 8, No. 3, July 2001 entitled "Einstein's Ether: F. Why did Einstein Come Back to the Ether?")

Chapter 2: Invisible Expanding Space

In this chapter we will explore the possibility of how our length standard might be changing relative to a reference outside the bounds of the Universe and still appear to be constant from our viewpoint. For these discussions, we will consider distance measurements to be static and instantaneously taken without considering time and the uncertainty principle. The principles of these discussions will apply to the traditional x,y,z coordinate 3D space, but for simplicity and ease in understanding, we will develop the concept in a planar x,y 2D space. We will consider 3D space later after these basic concepts are understood.

To begin, let's say that there is some kind of "fabric" to our Universe's vacuum space, (that space in which no matter can be seen). This fabric will be an ether of mass-less, motionless, invisible particles as Einstein described through which light can propagate and standards for space dimensions and time can exist. For discussion purposes, we will call them vacuum space particles (VSPs) with given and constant dimensions relative to our measurement standard. These particles are all similar in size and shape, but they are distinct and individual. The particles consist only of a volume of their own unique space material that cannot be shared with any other particle. They are all of the same size and symmetrical shape and are in intimate contact with each other with no "empty" spaces between them, requiring that they must be polyhedral and share flat common sides. Matter can exist and move around in the medium, displacing, compressing and decompressing the VSPs as they flow around the moving matter, but the particles themselves do not move. We cannot observe them directly, but we can see evidence of their existence through the actions they cause in observable matter.

For illustration in our 2D planar model, these particles would only occupy an area in a plane and would be polygonal in shape. There are a few polygon shapes that these particles could take on which would meet that VSP criteria, but let's select the simplest one, an equilateral triangle for our discussion. The VSPs may not actually be that shape as there could be other conditions that they must meet, but it doesn't matter for this discussion because the measurement principles we're discussing will apply to any consistent shape. They just have to be polygonal and symmetrical for 2D space (polyhedron for 3D space). Figure 2.1 depicts a section of 2D space consisting of equilateral triangles. Now our "ether" is "mass-less", invisible, and motionless, and consists of VSPs. They are the least dense particles in the Universe, so we will assign them a density of one. They will be used as a density reference for all matter in the Universe so that the density ratio of the matter to the space particle density is representative of the mass.

2D Space Particles as Equilateral Triangles

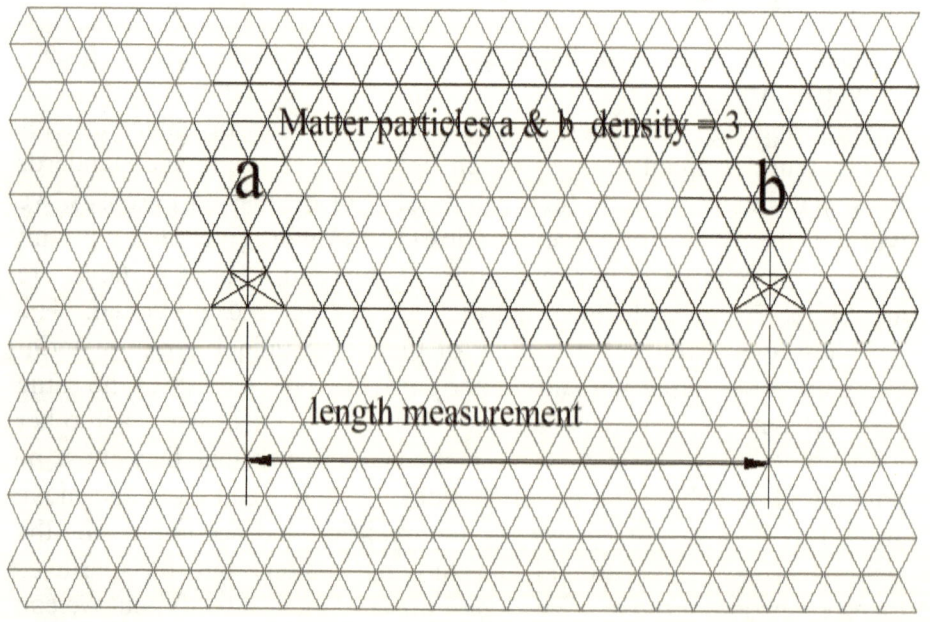

Figure 2.1

Figure 2.1 is a drawing representing our 2D space with equilateral triangle VSPs. These particles are the "fabric" of space through which distance measurements can be made and electromagnetic , gravitational, and inertial forces can exist. We cannot see or otherwise detect these particles, but that doesn't mean that they are nonexistent. We can still see the effects of their presence even if we can't see them.

Objects **a** and **b** are chunks of matter, (consists of clumped condensed space particles) have a density ratio of 3:1 (relative to the VSPs), and have properties of mass. Conventionally length measurements have been considered constant and unchanging so that the length of a physical reference standard would remain constant no matter what position it was in or where it was placed in space. So, we can take our "ruler" and measure the distance between objects **a** and **b** and also obtain the amount of space (or VSPs) between them. By doing this we have assumed unchanging consistency in the fabric of space relative to our measurement standard. Now might be a good time to mention that our physical measurement standard itself (the ruler) also contains a lot of "empty" space inside the atoms and molecules that it is made of. These spaces must also be filled with our triangular VSPs if we are to be consistent. With this in mind, we can define distance as a certain quantity of space particles along a straight line between two points. So the smallest measurable distance in the Universe could be no more than one space particle. The largest measurable distance would be the number of space particles existing between two points in the Universe's largest outer boundary dimension. Note that the "size"of a vacuum space particle in fraction of a meter terms relative to our meter reference is almost incomprehensible (if it could be seen), but it does exist and is finite nonetheless. Likewise, the amount of space particles in a straight line between the outer bounds of the Universe is probably even less comprehensible, but the quantity is still finite and not infinite.

So now in "empty" Universe space, a one meter span would contain a finite and constant number of VSPs stacked end to end to make up the one meter distance. Since this distance appears to be constant to us, we assume that this is consistent throughout all of space, even though we rely on light traveling through that very same space at a constant velocity to us to know of its existence (neither of which the consistency we are certain). We assume one meter of space a light year away from us would span the same distance on our measurement standard as it would here in our space. This implies that all space particles throughout the Universe have to be the same size and shape as our local reference in order for the distance to be consistent. And we assume that light speed is constant throughout space even though we don't know how it moves through it.

With this new way of looking at length measurement, what does it really mean to say that the distance between two objects is constant? It means that the *quantity* of VSPs, or the vacuum particle displacement between two points has remained constant. Also, if the distance between two objects of fixed size was measured, it means that the number of space particles the fixed objects occupy has remained constant as well. In summary we will state the definition of length or distance as we see it:

Distance between objects, or the measurements of the size of objects appears constant to us as long as the quantity of space particles between objects, or the quantity of space particles displaced by the objects remains constant to our measurement standard space particle displacement.

You might be thinking, so what? How does this make any difference in our understanding of length measurements or distance between objects in the Universe? Well there is one subtle thing about this that makes an enormous difference in our understanding of the *workings* of the Universe. It is possible that the space particles themselves could be expanding relative to the infinite non-universe void. As long as the particles and all objects of matter expand in the same proportions, they would appear to be constant and unchanging in size or in the distance between them. We could not see the particles or their expansion. This is tremendously significant! It means that it is possible for the whole Universe to be expanding from within. Each little particle of space could be generating new Universe space that cannot be shared with any other space particle, resulting in the entire Universe expanding into more of the non-universe space void without our detection. We could still be moving matter particles around in this expanding space medium, change distance between objects, etc. and the background space medium would still be invisible, and appear to be constant and unchanging to us (from a Newtonian mechanics perspective).

Supposing this expansion is the actual case, how would we measure things with our fixed ruler in the infinite 3D void space? In chapter one I introduced an "Infinity Quantization Ruler for a Finite Length Subset". This would be a handy tool to use when describing the unseen finite Universe expansion into the infinite Void.

Fig 2.2 represents such a ruler that has graduations on it for some fixed length standard. It looks like an ordinary yard stick that is finite and unchanging in our expanding Universe. But if it had graduations on it in a length standard that was constant in our Universe, and you were able to see it from outside our Universe in the infinite Void where the reference standard was constant there, this ruler would be stretching.

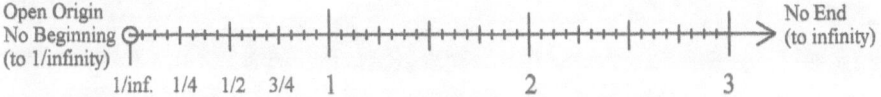

Infinity Quantization Ruler for a Finite Length Subset

Fig 2.2

If our Infinity Quantization Ruler had graduations on it in a length standard that was constant in the Void, and we were able to see it from inside our Universe, it would appear to be collapsing into an infinitesimal point. So, if we view the Universe from the Void with graduations on the infinite Void Cartesian axis in some constant length standard in the Void, we would see the Universe expanding everywhere like a foam.

What, you might ask, is a fixed length standard in the infinite Void that we can relate to in our Universe? Very good question! There is only one thing we can use, and that is the initial size of the Universe when it came into existence. If we assume that the Universe had a finite size at its beginning before expansion started, then we could use that length as an unchanging reference for our infinite void ruler and coordinate system. Logic says that the Universe is finite because it is expanding and not infinitely everywhere already. Also it is logical to assume that since it exists, is finite and expanding, it also must have had a beginning size when it came into existence. If we consider that each of the finite and inconceivably large number of space particles in the infant Universe existed in the beginning just before they started expanding, it would be very hard to imagine how small they would have been relative to our present expansion level. But, just remember that the infinite Void space has no limits in the small or large, and by definition, if something exists in the physical sense, then it must occupy space in the Void. Therefore the Universe and its individual space particles must have had an initial size.

Since we are going down this expanding Universe path, it would be good to point out here that this expansion is one way, that is, the Universe space that each particle generates from within cannot be made to go back where it came from.

Chapter 3: Universe Beginning, Expansion, Time, & Energy

Since it is possible that a Universe composed of expanding space particles could exist, let's explore how it would work and how all of the things we see happening in it could be explained with the actions of these particles. We will go for the grand unification prize and say that the entire Universe is composed solely of the space particles. It might be good to give you an overview of where I am going with this before I get into any details, so here goes:

- The entire Universe consists solely of expanding space particles. There are no voids in the Universe where these particles do not exist. The particle expansion comes from an unknown source within each individual particle.

- What scientists think of now as "vacuum space", areas of the Universe that contains no matter, consists of VSPs (vacuum space particles) that are the most highly expanded and the least restricted in their configuration with their surrounding space particles. These particles are assigned a density of one. They are used for comparison in the density ratio with grouped clumps of particles.

- All matter in the Universe consists of grouped clumps of condensed yet still expanding space particles. The mass of the clumps is proportional to the density ratio (with respect to VSPs). Large clumps like the nucleus of an atom may consist of multiple sub-clumps and even semi-expanded space particles within it.

- All of the invisible traits and attributes associated with the physical Universe; Energy, Time, Gravity, Inertia, Electromagnetic forces, ... are caused by the expansion of these space particles and their interaction with each other. There is nothing else in the physical Universe.

Matter Expanding?

A clumped group of lesser expanded space particles forms matter and has a density greater than one relative to the VSPs. These clumps are still expanding in the same proportion as the VSPs though, so that objects appear to be constant in size. The clumps may be in many different configurations and consist of various polyhedron shaped space particles. These clumps and configurations make up the various types of matter, and the density of the particles in the clumps give the matter a property of mass, the higher the density, the greater the mass.

Okay, now you are wondering what mechanism forms the clumps and how are they held together if the only thing in the Universe is space particles? Another good question, one that will take a bit of explaining! Remember that even though the space particles are identical, they are still uniquely individual and create their own brand of space that cannot be shared with any other particle. I will list a simple set of rules that expanding space particles must follow if the Universe is to consist solely of them:

1. Particles can only expand, they can never contract. The rate of expansion of space particles relative to a fixed reference outside the Universe can vary if the particle's expansion is restricted, but it will always be greater than zero. The expansive forces within the particles are equal, and nothing exists in the Universe to force any created space inside the space particles back to its source.

2. The space of one particle cannot be shared with another.

3. There are no voids between the particles. Particles terminate on each other with flat surfaces and facets of polyhedrons. Each facet of a space particle can only be shared with one other particle. Shapes of the particles can vary in size and number of facets to practically any voluminous flat shaped (vs. curved) configuration in solid geometry.

4. Since facets of adjoining space particles must be coincident, and particle space cannot be shared by space particles, they interfere with each others expansion forcefully. The more densely packed they are, the more they restrict each others expansion.

5. Since the particles must share sides with other particles, and there are expansion restrictions imposed by the boundary conditions with each other, groups of particles may clump together in all types of "Kaleidoscopic" configurations that are more dense than the VSPs.

So, to answer your question about what's holding the particles together, the clumps are being "pushed" together rather than being held together by some adhesive. *The only primary force in the Universe comes from the expansion of impenetrable space particles and their actions upon each other.*

The Beginning

How did the Universe begin? Well, no one knows for sure yet, or maybe that knowledge is not accessible from our vantage point, but here is one possible way it could have begun. In Chapter 2 we talked about the Universe having an initial size when it came into existence. It is finite even though it contains an enormously large number of infinitesimally small space particles. For our discussion we will assume that this enormous group of particles was spherical in shape (as it sat in an infinite non-Universe void), and that all particles were cubical and uniform in size. We don't know for sure if the expansion of each particle is continuous or comes in discrete sequential steps random to each other, but we will assume that the particles expand by random steps for now. At the onset of expansion, the particles in the outermost layer had one side or surface exposed to the Void which offers no resistance to its expansion. The only thing resisting expansion of these particles was its contact with other particles on the sides away from the Void. And so these outermost particles expanded in the direction of the Void at a greater rate, and that side of the space particle could have been curved instead of flat. This caused the particles to be highly irregular in shape, and since those particles were not expanding synchronously, the symmetry of the outer layer was broken. "Cracks" in the outer surface occurred, and fissures penetrating into deeper layers were formed as the particle's expansion restrictions changed due to the broken symmetry. These cracks and fissures were made up of oddly shaped space particles, but still they maintained the conditions of one facet in intimate contact with adjoining particles, no voids, etc. And so the infant Universe was wildly and chaotically fracturing and expanding at its beginning, but not from a "Big Bang" explosion, rather it was expanding everywhere from within.

Energy

Now you might be thinking, where's the energy that is "supposed" to be present as described in the super hot "soup" of the Big Bang theory? Well, therein lies the faulty assumption in that theory. With it, all of the energy in the Universe was present when it came into being, and all energy is conserved during the course of its existence. This is because the assumption is made that the energy is causing the expansion, as in an explosion.

Looking at this occurrence from an expanding particle perspective, which I will coin "The Big Fizz", the volume expansion is creating energy as it happens. There was no energy until the first particles started their incremental expansion. The infant Universe was super dense and at its minimum "size" in the Void. All of the particles in the inner layers were restricting each others expansion in a symmetrical way, and so they remained compact until the symmetry was broken. The particles in the outer layers expanded "rapidly", or it would seem that way to us because their expansion into the Void was much greater than the inner layer particles.

We are not able to see all of this absolute energy (so called since it is the total energy present in the Universe at any given level of expansion) because we can't see the expansion of the invisible space particles. The energy that we do see is the result of a rate of expansion increase in absolute space between two given quantities of particle space (more about this later).

You're probably thinking, Whoa! Wait a minute! What about the conservation of energy? Well, as it turns out, the energy that we see in our Universe is a result of a differential expansion between two volumes of space. If vacuum space and matter are expanding in the same proportions, the Universe energy being produced (that expansion which is creating it) cannot be observed because nothing we can see is in motion. If however a chunk of condensed space particles (matter) has an incremental change in its volume ratio with space, then we will see this as an exchange of energy. And as it turns out, the amount of energy that we see per incremental change in volume is constant, so to appear that energy is always conserved. But in fact, energy is just a product of independent volume expansion and therefore will always be sufficient and in equilibrium with Universe expansion.

Remember though that Universe volume is in the invisible third order units of absolute space (a quantity of volume in the Void), that is, in some fixed and unchanging units as viewed from the Void as described in Chapter 2.

Time

Now this might have crossed your mind; What about time? And at what "time" did the Universe begin? Well, this is something that might give you a touch of heartburn. Time is dependent on the linear expansion of space particles, and relative to its surroundings. That is, time has first order units of length as viewed from the Void. This means that time is not "universal" throughout the Universe, and is a local observance. This is because time is a record of sequential changes in linear expansion, moving forward and never backward. If expansion slows down in some part of the Universe relative to another, then there will be a corresponding slowing of time in that part relative to the other. Time is a record of changes in Universe linear expansion, and it began simultaneously with the first increment of expansion for that part. There was no such thing as time before the Universe began, and time did not start at zero. Time came into existence simultaneously with the Universe.

Since the cause of Universe expansion is from the space particles expanding, and they can never contract, time can only move forward or slow to a near stop. Time can never go backward for us at our present expansion level. Could we "go back in time" and visit parts of the Universe that aren't at our level of expansion or haven't reached our time in the Universe yet? Don't know, but suspect that we would have to stop our time and let that area's time catch up with us to get there or to see it.

Before I leave Chapter 3 and this introduction to the relationship between expansion, time, and energy, I would like to address the apparent contradiction of non-compressible space when you see Black Holes "suck" everything in and compact it. It is not a contradiction because it's our perspective that makes it seem that way. If you slow or stop expansion of a portion of space particles or matter, it will appear to be shrinking to us because we and the vacuum space around us are expanding at a faster rate.

Chapter 4: Universe Size and Age

The assertions that I made in the previous chapter about our concept of length, space, time, and energy will no doubt upset the apple cart, so to speak, in the scientific community. I expect rabid attacks on these statements, and probably some personal ridicule trying to disprove them with the conventional understanding of length as a basis of proof. But that will be of no avail here. The real work which has to be done, is to show that what we see happening in our Universe can or cannot be the result of the expansion of invisible space particles. This is a challenge to the skeptics and naysayers. I have some ideas of how this would work which I will discuss in the rest of this book, but it is not my intent to "prove" this. To do so will probably take many intellectuals of much higher caliber than me working on it in this century for as many hours as was spent working on the Big Bang theory in the last. My approach is, I'm convinced that this is how the Universe exists and if it really is true, then there must be a way to show how it could work like this. In my Hippie friends terms, " I am all positive energy here". If successful, we will have found the illusive Holy Grail of cosmology: The Grand Unification Theory.

Length in the Void

How do we convert our Universe length units into the Void absolute length units? We will need this to "see" the Universe expand in the Void. Since we are expanding with time, or rather, time happens linearly with expansion, the Universe length units in the Void absolute length units is ever increasing with the Universe's expansion. Now we might have to use just a smidgen of math to better define this (my apologies). We will let **La** represent absolute Void length in some fixed and unchanging units. As I mentioned in chapter 2, about the only thing we can relate to length wise in the Void is the initial size of the Universe when it came into existence. So we will let this be the case, and we will call this unit a *Princh* (contraction for primary inch). Our finite measurement units in the infinite Void scale then, will be a ratio of current Universe size to its initial size as the Universe expands in the Void. And we will let **Lu** represent our standard and unchanging length units (meters) in the Universe. Also, we will let **Ts** represent our Universe time in seconds since it began. We will also need **K,** a constant of proportionality conversion factor, to adjust the result so the units are in the right proportions. **K** has units of Prinches per Second per Meter, that is $La/(Ts \cdot Lu)$, so that the product will be in units of absolute length. With these definitions we can express how our length would be observed from the Void by the formula:

Equation 4.1 $\qquad La = K \cdot Lu \cdot Ts$ *Prinches*

Space-Time

Notice that the La/Ts portion of the constant **K** represents the amount of Universe linear expansion per second in the Void with our time passing! Having a rate like this implies that the expansion is a function of independent time. But this is not the case here. *The only independent variable in the entire Universe is the continuing creation of Universe space volume from within each of the space particles.* This growth is the cause of Universe existence. As the particles grow in size linearly, time passes. (Linear expansion is in one direction along a line. Now there is also 3D volume expansion going on with linear expansion which equates to energy, but that is a topic for later discussion.) I will plant the seed here that the linear expansion rate is what causes the limiting speed of light, and will talk more about it later as well. Remember that space particle expansion is not observable as space expansion in the Universe and comes from within each of the space particles, sort of like dough rising. It can only be seen in our minds as expansion of Universe space in princh units in the Void. From our perspective here in the Universe, we can only see the result of linear expansion as time passing.

Equation 4.1 would be a simple expression if the time that we see in our Universe was independent and uniform throughout. But it is not. Any time that we see passing in our Universe is the result of an *average* linear expansion of the particles in a *selected volume* of the Universe with respect to the *viewing volume* average linear expansion. This is the connection to and relativity between time and space.

Okay then, what about our view of the Universe on a larger scale such as volumes containing galaxies in our "local" area? Well this gets real "sticky" about now, as in we have to "view" these galaxies through an expanding vacuum space. How are we even able to view these objects from afar in the first place? Here's where I'm forced to introduce how energy and forces can travel through vacuum space and affect distant objects with expanding space particles.

Remember that one of the rules of a space particle Universe is that it consists solely of expanding polyhedron shaped space particles with adjoining sides that must be shared with one another, and there are no voids anywhere in Universe space. The non sharing space of each particle forms the boundary surfaces at the polyhedron sides. And since these surfaces are the places where the particles are in intimate contact with each other, any lateral expansion on those surfaces have to be shared by both particles. If two undisturbed expanding vacuum space particles of identical configuration and size are sharing a common facet, and for some reason one of the particles is disturbed or distorted in expansion, then the adjoining space particle will be distorted in expansion on that side as well. This disturbance will then propagate throughout the vacuum space, one particle to the other, in the form of expansion ripples at the speed of particle expansion. This propagation speed, the particle expansion speed, is what we see as the speed of light. Any matter (clumps of compacted space particles) within the vacuum space will be expanding inversely proportional to its density on average, so that the length ratio with space expansion remains constant.

Now back to Equation 4.1 and the complication of knowing how big the Universe is and how long it has been in existence. It is possible that portions of Universe vacuum space outside of our local area may vary in density with respect to the vacuum space in our locality. In that case the expansion rate, the speed of light traveling through it, and time in that space would be different than ours (not happening at the same rate). It would be impossible for us to know this though since we depend upon seeing the light energy traveling through that same space which has a different speed of light. Since the *relationship* between space expansion and time in that space *would be the same* as in ours, we would not be able to detect anything different.

Back in chapter 3, I introduced a scenario in which the Universe could have begun. The cracks and fissures in the infant Universe's expansion happened chaotically and spread deeper and deeper into the compact interior. Some clumps of less dense space particles were formed during this process. And in some areas, the particles broke into freer expanding rudimentary VSPs (vacuum space particles). If our consciousness could have been there to see it, it would have been a tremendously spectacular display of energy as the space particles expanded into the Void. This expansion could have been easily seen in the Void where the length measurement comparison was constant against the initial size of the Universe. But from our view inside the Universe, the space particle expansion sizes and rates at this point in its existence would have been chaotic and random due to the irregularity in the growth of the cracks and fissures. For those inside parts of the Universe where the expansion was slower or had not started, time was passing slowly or had not started, and no actions of the Universe were displayed for us to see except for gravity, which is for later discussion.

When astronomers look out into deep space, they are everywhere looking at the outer layers of the expanding Universe shell via expanding space ripples in the vacuum space (electromagnetic energy). The outermost layers, most likely will always be expanding faster than the layer we are in because there is less expansion restriction in Universe space as the Void is approached. In Chapter Two, I made a statement that the distance between objects or the size of objects appear constant to us as long as the quantity of space particles between objects, or the quantity of our vacuum space particles displaced by the objects remain constant. As the distant matter and vacuum space become less dense with respect to our space and VSP references, it appears to us that the visible Universe is expanding everywhere we look. This expansion we can see is the so called Hubble expansion after the name of the man who is responsible for discovering it. Also, the incremental volume expansion occurring in those regions equate to a release of electromagnetic energy that can propagate through the vacuum space. When we see these volume expansion ripples come to us from afar in quantum steps at some rate relative to our expansion and time passing, we see these as a visible light frequency of electromagnetic energy. These frequencies can be "Doppler shifted" with movement just like the pitch of sound waves from a moving vehicle shifts as it passes by. Some of the light energy released from those far regions has been Doppler shifted down to a microwave frequency that is observed as the Background Radiation anywhere we look out into space.

Time is and has been passing quicker in the outer shells of the Universe, and the speed of light is greater there. When Cosmologists estimate the age of the Universe using the Background Radiation frequency and Hubble expansion rate observations with our constant Universe length and time measurement units, and assuming a constant speed of light, they see discrepancies because our "constants" are not really constant in the big scheme of things. These discrepancies lead to the invention of things like "The Great Attractor", "Dark Matter", and "Dark Energy" to explain the behavior. But, if we consider what we observe with an expanding space particle perspective, we can see that these correction factors are not needed to explain the behavior, even though we haven't determined how much we have expanded since the Universe began.

Chapter 5: Cause of Gravity and Inertia

In Chapters 3 and 4, I introduced the expanding space particle as the sole substance of the Universe. The only thing distinguishing matter in the Universe is its property of mass, which is related to the density of the matter relative to the VSPs. I listed a small set of rules that these particles must follow to be the basis for Universe existence, and I suggested how these particles might behave to exhibit the time and energy properties that we see.

Measurements using units of Mass, Length, and Time as defined in the Universe are what Scientists have traditionally used to observe and describe the workings of the Universe in the Big Bang (BB) model. As we have seen though, time and length measurements both depend upon the expansion of particles to exist. The Absolute Energy of the Universe is the result of volume expansion of each of the primary particles into the Void which we cannot see. Because this energy is invisible, it is assumed that it doesn't exist; it is not in the BB model. The energy that we do see and that is accounted for in the model, is the differential volume expansion between two volumes of Universe space. We will not be able to use the traditional Mass, Length, and Time measurements to describe actions of the Universe in the Big Fizz (BF) model as was used in the BB model. We can however translate the actions predicted in the BF model using its terms, into terms that we would see in the BB model. By doing this we will have verified the accuracy of the model by actions we can observe. This will be the approach taken in our quest for knowledge about the Universe's existence.

I have been giving you "tastes" of various aspects of this BF theory without going into much detail so that you will be somewhat acclimated to and interested in these things before getting "knocked down" by the "strange and crazy" ideas that I present. I have coarsely shown how it is possible for these invisible space particles to exist and to cause some of the actions we see, but the "Devil" will be in the details of connecting all that we see to them. This is the work that has to be done and will require many hours of intellectual labor sorting it all out.

Particle Interface Reactions and Considerations

If all that exists in the physical Universe is expanding space particles, then there must be certain behavioral rules that space particles follow based upon their simple nature. We assume here that: There are a finite number of particles and no more of them are being made, that the particles are individual and they cannot share their space with others, that each particle contains a source of expansion space, and that the expansive forces within them are equal. We assume that entire Universe consists of individual particle space, and no voids can exist between the particles.

Each particle is trying to forcefully increase its volume against the resistance of its neighboring particles which have an equal opposing force. It follows that the surface boundaries adjoining the particles must then have flat sides (if the forces are equal within the particles, then the boundaries can't favor one particle or the other with curved surfaces), and this requires that the particles must be polyhedrons in shape. It is logical to assume that since the particle's expansion is restricted at its boundaries, it would take on a shape that would maximize the volume to surface area ratio within the limits of other restrictions placed upon it. We know and can see that there are some denser groups of particles that make up matter with mass, and that there must be something that causes them to remain compact. About the only things possible that can control expansion of the particles is their geometric configurations and a surface interface condition that requires solitary congruence between sides of interconnecting polyhedrons. With this assumption then, how can a dense group of space particles (matter) move through a stationary VSP medium and still maintain congruence with the stationary VSP particle's sides? Does the VSP side that is in contact with the smaller matter space particles "stretch" as it moves and then gradually "roll" into a different polyhedral configuration with more sides to accommodate the additional sides as it passes through, and then "peel" away as it leaves the VSP? Understanding the mechanics of this movement will be required before we can fully understand the space particle's nature, but we can still move forward with our knowledge on interaction without knowing exact details of the movement.

Establishing exactly how the different geometrical configurations with congruent faces control expansion will no doubt require a great deal of work. The only way I know how to determine these things is by trial and error. That is, initially assume certain things are true by logic, and then proceed to show by modeling and deductive reasoning that these things can or cannot cause the behavior we observe. We have little in the way of observing and modeling the dynamics of an expanding non-compressible fluid, especially when the fluid is made of particles that assume polyhedral shapes with congruent sides. But unlike the nineteenth century physicists, we now have 2D and 3D computer modeling tools through which we can virtually see how hypothetical expanding particles with certain conditions put upon them would behave. Using this tool as a means viewing the unseen, is just as valid as using a telescope or a particle accelerator to see things we otherwise cannot. Through it we can determine the validity of our assumptions.

Gravity and Inertia

In this chapter we will discuss how gravity and inertia could work in this BF model. Even though these properties of mass are two separate things, they are oftentimes lumped together and defined only by their behavior in mathematical terms without a physical cause. This would be okay if we want to believe that these properties are rudimentary and an inherent nature of matter without any physical cause. But this is not acceptable for most of us. Even Issac Newton, who first observed and formulated the gravitational and inertial behavior with mathematics, didn't want to believe that objects could affect each other at a distance without some physical mechanism causing it. Probably the main reason that it disturbs us is because without knowing the cause of gravity and inertia and its connection to the physical mass, we can't be for sure that we will know the behavior in all situations. The mathematics is not a cause, and if for some reason our model is wrong or incomplete, then we really can't predict its behavior or know how it works. This is what happened when Albert Einstein discovered the relationship between space, time, velocity of matter, energy, and the speed of light. We found that Newton's laws of motion and gravity were only a partial model of the behavior. A new mathematical model had to be developed that would work for Newtonian mechanics and the Special Theory of Relativity. But still, the mathematics is just a mind tool that helps us to understand, and not a cause for anything physical. Albert, like Issac, believed that there must be something else related to the physical that is causing the gravity and inertial behavior.

My goal in the following section is to show the physical connection of these properties of matter to the actions of the invisible expanding space particles permeating the Universe. *These space particles, though invisible, are physical and real, are the only substance of the Universe, and their expansion causes all actions in it.*

Vacuum Space Revisited

Chapter 2, Figure 2.1 is a simplified illustration of a hypothetical planar vacuum space area containing particles of matter. This figure is sufficient to show how measurements are made, and how they are relative to the vacuum space in which they are made. But in order to show how gravity and inertia works in a 3D Universe vacuum space, we might have to complicate the model some.

In Chapter 3, I presented a set of rules for the space particles. Additional requirements are needed for particles in vacuum space (Universe space without matter). VSPs (vacuum space particles) must be uniform in size and shape and be expanding at the same rate relative to each other. These particles must also be symmetrical so that orientation is not a factor when discussing actions occurring within vacuum space.

In order for Vacuum Space Particles to be symmetrical, of the same size and shape, and to share any one of its polyhedral sides with only one other, the sides of the polyhedron must also be of the same size and shape. And since there can be no void spaces between the particles, every facet of the polyhedron must be in full and intimate contact with another particle's facet. There are only a few possible polyhedral shapes that could meet all of these conditions.

Again, working in a simpler 2D planar space for illustration purposes, the equivalent planar shapes would be the Equilateral Triangle, Square, and the Hexagon. Figure 5.1 illustrates that these three regular polygons are the only shapes that will meet the 2D VSP requirements; Equal side lengths and interior angles, symmetry, sides coincident with only one other polygon side, and no gaps or overlaps between particles.

For a 3D Universe model, the Regular Polyhedrons are called Platonic solids, and these will be explored in the next chapter when discussing electric and magnetic forces. But for now we will continue to use the 2D model to illustrate gravitational and inertial forces. At this point in the development of the 2D Big Fizz theory, we have no way of knowing which one of the three possible shapes the VSPs could take, but I will choose the hexagon. Of the three shapes, the hexagon has the

Possible Shapes for 2D Vacuum Space Particles

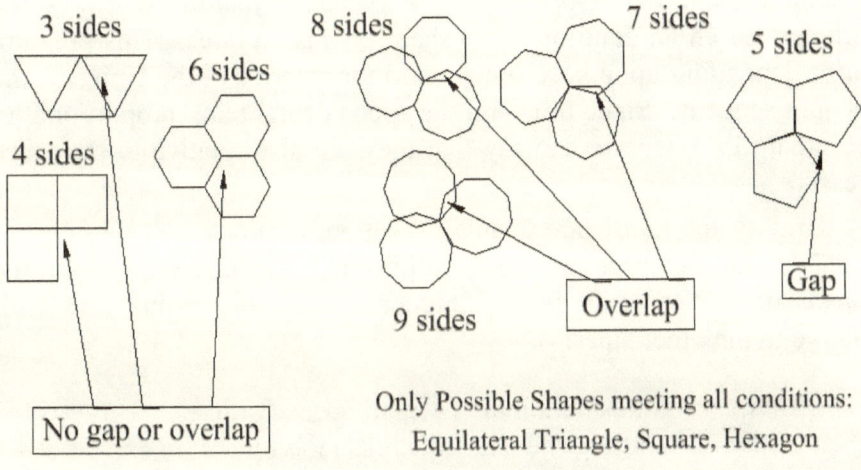

Only Possible Shapes meeting all conditions:
Equilateral Triangle, Square, Hexagon

Figure 5.1

highest ratio of area to perimeter length. And since each particle is trying to expand to its maximum area against restrictions on its perimeter, it will seek the shape that gives it the maximum area for a given perimeter.

Gravity

Figure 5.2 depicts a region of Vacuum space with two particles of matter (**M1** and **M2**) in it. This illustration is highly simplified to demonstrate the effect. In reality, these bits of matter would be much larger relative to the space particles, and consist of a composite of multiple clusters of clumps and sub-clumps held together by the restriction of expansion that the clumps have on one another. However, this illustration will be sufficient to gain a concept of how gravity and inertia works. Looking at this static diagram, it's easy to forget that Vacuum Space is actually like an non-compressible expanding fluid, and that in reality, this diagram is expanding at the "speed of light" in the Void. We have had very little, if any experience with this kind of fluid dynamic physics (this would be an exciting project for one of you for one of you brilliant budding Physicists).

The rate of flow of new space material from within each particle and its resulting expansion, is limited by the surrounding particles and their expansion. Since we are located somewhere inside of the expanding Universe and not on the Void boundaries, there is something akin to a *vacuum space pressure* that develops as the particles reach an equilibrium in their expansion rates. This pressure varies depending upon how "deep" into the interior of the Universe the vacuum space is. Since time and the speed of light are proportional to the expansion rate, we can say that they are also relative to the space pressure.

Gravity and Inertia are related in that both actions are caused by an imbalance of the space pressure equilibrium. The forces acting on the objects of matter result from a higher pressure on one side of it relative to the other sides.

Expansion Restriction of Vacuum Space Particles,
Distortion, and Consequential Gravity

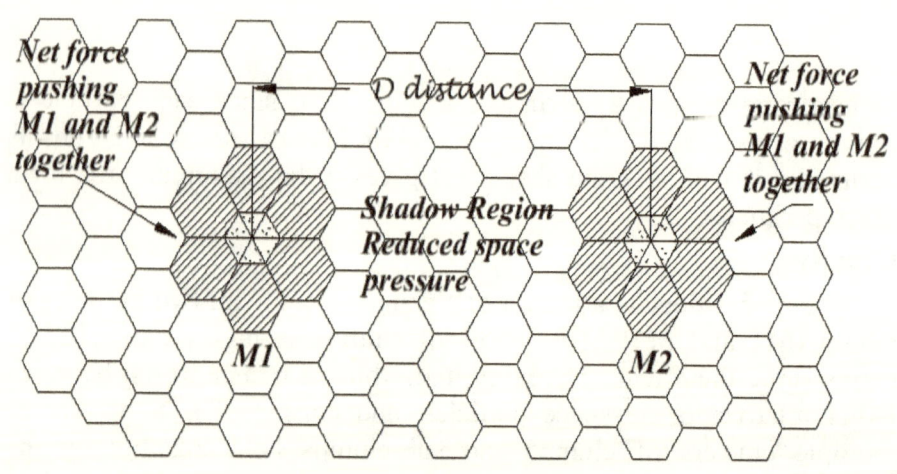

$$Force = G\,(M1 \cdot M2)/D^2$$

Figure 5.2

In Figure 5.2, each object is a bit of matter represented by six compacted space particles per volume of vacuum space particle. The mass of each object is 6. The vacuum space particles surrounding the bits of matter (shown as shaded) are distorted due to shorter sides of the matter bits that are in contact with them (remember that in all cases

46

adjoining sides of space particles must be coincident). This distortion in expansion causes a slight reduction in the space pressure around the matter because the affected vacuum space particles can't expand as fast with one side restricted.

In practice, this distortion around the matter is not limited to just the vacuum space particles the matter is in contact with. It passes on to other particles in the region, one to the other, diminishing in distortion as it moves away from the cause. If the object is isolated in space from any other object, then the net force on the object due to the space pressure is zero because the forces acting on it are equal all around. If however another object is brought into its vicinity, the average space pressure between the two objects along distance D will be reduced due to the addition of the two reduced pressures surrounding the objects. This reduction in pressure causes a net force acting to push the two objects together. The closer the objects are to each other, the greater the reduction in space pressure between them. This is because a bigger portion of the overall average pressure between the objects is composed of the reduced pressure region around the objects when they are closer together. This reduction in average space pressure and resulting increased force as the objects come closer together is responsible for the gravitational behavior law described by Newton:

$$F_g = G \cdot M1 \cdot M2 / D^2$$

Space Distortion and Constancy of Gravitational Laws

Scientists talk about "depressions" in the fabric of space around a highly massive object. This is in fact just the distortion of the vacuum space particles surrounding the object. This distortion severely warps the space particles out of their natural shape, distorts length measurements, and alters the straight path of any electromagnetic energy wave that would be traveling through it. The same thing happens in smaller less massive objects, but the distortion is not as severe and the layer is much thinner, so it is not normally detectable.

What happens to the gravitational force as the Void boundary of the Universe is approached? Well, since the gravitational force is the result of an imbalance in space pressure upon the objects, any variations in space pressure would also affect the gravitational force. Just speculating here, but points in the outer regions of space close to the Void probably would have little restriction in VSP expansion. This would result in a lower space pressure. With this reduced pressure, the magnitude of the gravitational force upon objects is probably not the same as it would be in higher space pressure regions. Although not precisely defined yet (another good project for someone), we can say that the gravitational force between objects and through all distances apart is not constant and unchanging throughout the Universe because of this.

Since traditionally it has been assumed that the gravitational laws we have formulated were everywhere constant, using length measurements that were assumed to be constant throughout the Universe, it is not surprising that there are discrepancies in the Big Bang model that require things like "Dark Matter" and "Dark Energy" to make it work with what we see happening in the far reaches of the Universe.

Inertia

Since the Void is infinite for our purposes, there is no "place" in it for a location reference. There is no particular place for the center of the Universe to reside either. If there was, it would always be moving as the Universe expanded anyway. As each space particle expands from within, everything expands and the center keeps moving relative to a fixed space outside of it. But since we are observing the actions of the Universe from inside it, we can use the Universe itself as a reference.

The vacuum space particles, even though they are expanding, appear stationary to us. Objects with mass can move in position relative to them. When an object with mass (high density cluster of space particles) moves in the stationary vacuum space particle medium, it causes the VSPs against the direction of movement to be restricted in their expansion. The particles resist by applying a force to the object that is opposite the direction of movement because of the increased space pressure on that side. When an object is accelerated in vacuum space, this force is seen as the acceleration force, the force first described mathematically by Issac Newton:

$$Force = mass \cdot acceleration$$

Okay then! But you might be wondering, what keeps the particles from continuing to apply a resistive force to the object when it is not being accelerated? Remember that other inertial law of motion: "*A body in motion tends to stay in motion unless acted upon by an outside force*"? Why isn't the force of the restriction of vacuum space particles

48

slowing down the moving object? The answer is, when the vacuum space particles flow around the moving object, they give the object a "kick in the butt" as the object passes them and they expand back to their normal shape. When an object is moving at a constant speed, the quantity of particles and their degree of expansion restriction in front of the moving object, is equal to the quantity of expanding particles behind it. Therefore there is no net force on the object due to the movement. If however the velocity of the object is changed at some rate (accelerated), the degree of restriction in the direction of the acceleration will be greater or less (depending on the direction of acceleration) than the expansion behind it, causing the inertial acceleration force to be felt on the object.

I have deliberately not mentioned how an external force can be applied to an object to move it in space, and have highly simplified the illustrations in this chapter so that the functional concepts of the gravitational and inertial forces could be understood easily. All of the physical objects that we normally encounter and can see with our own eyes, like a rock for instance, is a complex cluster of sub-clusters and clumps of space particles that make up the elemental atoms and molecules of matter. The components of atoms, the electrons, protons, neutrons, etc., and even the subatomic particles, quarks, etc., are still just space particle groups of different geometrical configurations, even though these micro-clusters have to be treated differently when trying to observe them (more about this later). The only thing holding them together is the restriction of expansion they have on each other imposed by the sharing of sides and inability to share space.

The vacuum space particles are the only "non-clustered" and least dense particles in the Universe. They are the ones that have the maximum volume (as compared to matter space particles) and are symmetrical and regular (all sides and angles equal). Clumps of space particles that form matter with mass, can take on practically any solid geometric shape that has flat sides and volume. The number of sides the particle shape has could vary, and it could be highly irregular, but the one side must be coincident with only one other particle side requirement still holds. It is this requirement that causes the space particles of the object with mass to distort the vacuum space particles around it and reduce the space pressure in its vicinity. Also, any change in the shape of the particles that results in a change of the volume to surface area ratio will be exhibited as an energy release in our Universe. Any changes in the shape of the particles that affect the overall expansion rate of a given volume of space will also affect the passing of time in that space.

Chapter 6: Electrical, Magnetic, & Gravitational Fields

I enter into this chapter with somewhat of a dread. For you nonscientific folks, this chapter might be kind of tough to understand, but don't get discouraged. Everyone should be able to get something out of it. Just wade on through the math and glean whatever you can gain from it. You don't have to fully understand it. The main thing I'm trying to show here, is how the expanding space particles cause all of the invisible "magic" and mysterious forces that we see working in our everyday lives.

Trying to understand the behavior of space particles such that they produce electrical and magnetic forces seems to be an ominous task to me. Like gravity and inertia, electricity and magnetism are connected, but they have a more complicated dynamic connection in which one force can produce the other force in an orthogonal plane (planes at 90 degrees to each other) with movement. Here we will have to use 3D space, and this will likely be harder to physically model, so we will take it rather slowly. We will talk about static electricity and the things we know about its behavior first, and how the space particles must behave to produce the results. Here is a list of things we know so far:

1. There are two types of electrical charges; positive, and negative.

2. We know that the charges come in discrete quantities and are confined to matter; objects that have mass.

3. We know that unlike charges (a positive charge and a negative charge) exhibit an attractive force between them, and that like charges (positive and positive, or negative and negative) exhibit a repulsive force between them.

4. We know that electrical forces between charged objects can be transferred through the VSPs (vacuum space particles), and that the forces are transferred at "the speed of light".

5. From previous chapters we know that the only forces that can be transferred through the VSPs to objects is caused by an imbalance in their expansion, and that the VSPs remain stationary relative to the Universe while expanding.

6. We know that a charged object with mass, such as a proton or an electron, can move through the stationary VSP medium, or

remain stationary relative to it, and simultaneously exhibit properties of mass and properties of charged objects.

In a static situation (charged objects that are not moving) the amount of charge an object has with reference to the electrical forces is equivalent to the amount of mass an object has with respect to gravitational forces. That is, the greater the amount of charge, the greater the electrical force upon the charged objects. As a matter of fact, the formulas describing the behavior of the forces between two objects with charge or mass is similar. I apologize for having to use math formulas for you nonscientific folks, but I will use them here to show the similarity:

$$Gravity\ force = Fg = G \cdot (M1 \cdot M2) / D^2$$

$$Electrical\ force = Fe = K \cdot (Q1 \cdot Q2) / D^2$$

where **M1** and **M2** are the mass of the objects, **Q1** and **Q2** is the amount of charge on the objects, **G** is a gravitational constant of proportionality, **K** is an electrical constant of proportionality, and **D** is the distance separating the two objects.

It is probably more than just a coincidence that these two separate forces are alike in form. I believe that is because the forces are transferred through vacuum space by the same means; through the change in the "normal" expansion of the stationary VSPs passed one to the other at the speed of their expansion (speed of light). From previous discussions on gravity, we know that the only forces that can be "felt" on an object is either caused by an imbalance in the space pressure around the object, or that applied by the movement of another object against it. Since no other objects are in motion around the charged objects in a static electricity situation, the electrical force transferred to them in vacuum space must come from the space pressure differential around them.

The VSPs around the charged objects are distorted when they are around the massive objects, but the electrical distortion shape is different from the gravitational distortion shape because it does not directly affect the gravity distortion. Also, the positive and negatively charged objects must have VSP distortion shapes that are complimentary if the objects are to have an attractive force imparted to them from the VSPs, and for like distortions to have repulsive forces imparted. As we will see, there must be an additional possible static magnetic distortion in the symmetry of the VSPs that is different from both the gravity and electrical distortion that does not directly affect either of them.

Here's a bit of logic to consider: Since we have assumed that the Universe consists entirely of expanding space particles, and that their expansion and interaction with each other is responsible for all observable actions in the Universe, then in 3D space there are only three orthogonal dimensions or degrees of freedom in which the VSPs or other space particles can expand or be restricted. If we consider each VSP individually, we could establish the positive space occupied by the particle in the Void with the origin of an x,y,z coordinate system centered on it, and with positive axes in all directions (there is no negative space). The measurement unit scale could be in Prinches (based on the initial size of the universe).

VSP Coordinate System

Coordinate system for a VSP in the Void. Unit of measurement is in Prinches. These units are based on the initial size of the Universe when it came into existence at the first increment of expansion.

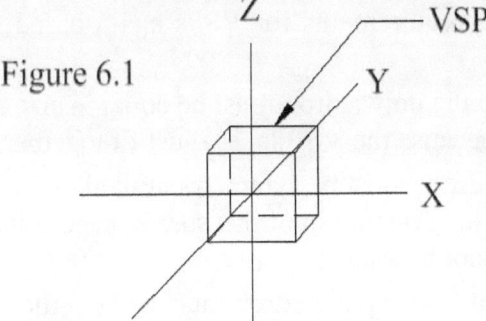

Figure 6.1

Origin of the axes are centered on each VSP in the Void.
The center of the Universe is everywhere within itself in the Void. The Void is infinite and has no reference or origin, so the Universe itself becomes the location reference for our purposes.

In this view from the Void, the center of the universe is everywhere within itself, and in no definable place in the Void (because the Void has no defining origin). Since we know that the Universe is expanding, and the expansion is the cause of its existence, then we know that the Universe must be finite or have a finite set of space particles, because it is not everywhere already. In this sense, and from our viewpoint inside the Universe, there exists a centroid (middle point) for it. But just remember as we talked about in Chapters 3 and 4, our length measurement standard (the meter) is not constant throughout the universe, so it might be impossible to know where the middle point is.

In a static situation, the only known forces between objects with mass that can be transferred through the VSPs in 3D space are the gravitational, electrical, and magnetic forces. And in order for the VSPs to do this through restricted or expansive distortion passed one to the other without interaction between forces, the distortions have to be orthogonal or at 90 degree angles to each other. (An example of this is a rectangular box in which you could make taller or wider or longer without affecting any of the other dimensions). This is fortunate because we have just enough directions in the x,y,z coordinate system centered on each VSP to accommodate the expansion for the three forces... or maybe it's not a coincidence!

VSP Shape

In 3D space, what polyhedral shape does the VSP take? What polyhedrons meet the requirements for a VSP in vacuum space? Let's list the polyhedron requirements for VSPs again based upon what we know so far:

1. All faces of the polyhedron must be equal in size and shape. This is so because the VSP faces must fit together perfectly.
2. The polyhedrons must be symmetrical in all directions. This is so because the orientation of measurements or direction of forces will not be a factor.
3. Each face of a VSP polyhedron must be congruent with one other face of an adjoining VSP, and no gaps or overlaps can exist between VSPs. This is so because no voids are allowed between VSPs.

Now at first it might seem that there could be many polyhedrons that would meet these requirements, but upon closer examination we find that there are only five polyhedrons that meet the first two requirements for the VSP shape. These are the Platonic Solids, and were proven to be the only polyhedrons with equivalent faces composed of congruent convex regular polygons by none other than

Euclid himself. The Platonic solids are: the Tetrahedron, Cube, Octahedron, Dodecahedron, and the Icosahedron. The ancient Greeks knew of these and were described by Plato 350 years before Jesus walked the earth. Even a thousand years before that, the neolithic people of Scotland developed stone models of the five solids, and they still exist in a museum today. Strangely enough, they are sometimes called the cosmic figures.

Platonic Solids

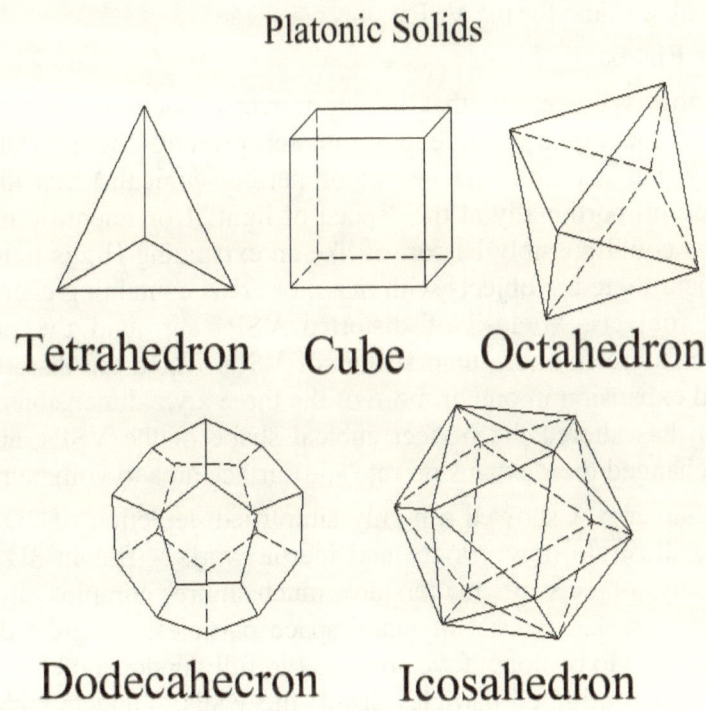

Tetrahedron   Cube   Octahedron

Dodecahecron   Icosahedron

Back to the shape of the VSPs after this historical digression. The VSPs must also fit together perfectly with all facets congruent upon only one adjoining VSP's facet. So we might think there may be two or three possible Platonic shapes that would meet this requirement for 3D space, as there were for the regular polygons in 2D space. The most obvious shape that will meet all of the VSP shape requirements is the Cube. As any child who has played with blocks knows, cubes can be stacked together to make a larger solid block without any gaps or irregularities. But what about the Tetrahedron? This is not so obvious, and we would have to do a geometrical analysis to find out. I will spare you nonscientific readers and not go into these details here (more in the appendix), but it turns out that the Tetrahedron won't fit together all around without an approximate 1.5 degree gap between sides. This is not a lot, but there can be no gaps, so this shape is out. A similar analysis on the Octahedron comes up with the same result; an approximate 1.5 degree gap on the sides. The analysis of the Icosahedron and the Dodecahedron are more complex, but they will not fit together perfectly to make a solid either. This leaves the Cube as the only possible shape for the VSPs!

Invisible Fields

So now we can say that in our universe vacuum space where objects or electromagnetic energy is not present, there exists this stationary and invisible 3D matrix of perfect cubical VSPs that are expanding proportionally at the "speed of light" (for scientific minded folks, this could possibly be sort of like an expanding Higgs field). In areas where there are objects with mass, or areas containing charged or magnetic objects, "fields" of distorted VSPs surround the objects. These fields are no more than strings of VSPs which are distorted by restricted expansion in one or more of the three x,y,z dimensions. This distortion has altered the perfect cubical shape of the VSPs, and has slightly changed their expansion rate and surface area to volume ratio.

In chapter 5, I showed a highly simplified depiction of 2D space matter to illustrate how gravity and inertia work. But in 3D space reality, any objects of matter are much more complex in their configurations, and consist of many space particles. A great deal of work will have to be done if we are going to fully understand how these grouped and configured particles distort the VSPs. Objects with mass contain many space particles, so the act of distorting surrounding VSPs must be a cumulative effect from the space particle configurations rather than that shown in the simplified figure 5.2. That diagram is sufficient to gain a concept of how gravity and inertia work, but we will need to develop the model a little further if we are going to show how the electromagnetic forces are applied from VSP distortion. A situation similar to objects of matter and VSP distortion exists for the electric and magnetic forces affecting charged objects. Any object of matter considered here is a composite of clumps and sub-clumps of space particles that is much larger than a single VSP at any given instant.

The smallest unit of charge is that of an electron. The electron has a negative charge, but positrons do exist which have a positive unit charge equivalent to the negative unit charge on the electron. The mass of the electron and the positron at rest (apparent mass changes when moving) is 9.11×10^{-31} Kg. So these objects with charge must contain the same amount of configured space particles, just that the configurations are complementary to produce the opposite charges. Also, since these objects with a single charge are composed of many space particles, the distortion imparted to the surrounding VSPs must be a cumulative effect from the space particle configurations comprising the charged object. Likewise for the proton which has the same charge as the positron but has a much higher mass.

Figure 6.3 depicts non-distorted expansion, and the three distortions possible for VSPs surrounding an object of matter. Note that these diagrams are of the proportional expansion rate of the VSPs in three orthogonal directions, and therefore all axes are positive in both directions (contractions of created space in vacuum particles does not exist in the Universe). We could distinguish the negative direction as X',Y',Z' if necessary, but the distance would still remain a positive value . The origin of the X,Y,Z coordinate system is centered on a particular stationary VSP so that we can visualize how that VSP is affected by the presence of matter in its vicinity.

Remember that the non-distorted VSPs are cubical in shape, stationary, and symmetrical all around. So, it doesn't matter whether or not the orthogonal axes of the cube (as described and shown in figure 6.1) is aligned with the expansion axes shown here in these diagrams. These diagrams show the proportional expansion rate in any arbitrary orientation of the VSP axes and are *aligned with the object of matter* that causes the distortion.

Proportional Expansion

Now there is likely to be some confusion when we are talking about an arbitrary object expanding proportionally, so I'll expound on this a bit here before going into the expansion distortions.

Most of the time we think of expansion in terms of a ratio. We might say that baby Tommy has doubled his size since he was born (ratio of 2). If he was 16 inches long when he was born, he would now have grown 16 more inches for a total length of 32 inches. What about his width? If he was 5 inches wide when he was born, would he have grown 16 inches in width? No, he would have grown five more inches in width to twice his birth width, or 10 inches wide. This way Tommy maintains his shape and proportionality.

When we talk about expanding in all directions the same amount, then we are talking about *linear* expansion. When we are talking about all dimensions expanding in the same ratio, then we are talking about *proportional* expansion. In proportional expansion, all of the dimensions grow by the same ratio as they expand. For example: If a rectangular box that was 6" long by 4" wide by 2" tall expanded to twice its size, the new dimensions would be 12" by 8" wide by 4"tall. The amount of expansion in each direction is not equal because the original dimensions were not all equal. If a cube that was 4" long on a side doubled its size, then all of the sides would expand equally by 4" to make an 8" on a side cube. In this case the sides would expand linearly by the same amount, but if you measured the expansion across the diagonal between two corners of the cube, you would see that the cube expanded linearly by 6.93" instead of 4" in that direction. So, as it turns out, the only object that expands linearly the same in all directions with proportional expansion, is a sphere. Now back to the diagrams.

The diagrams in figures 6.3(a-d) represent the *proportional expansion* of a cubical VSP. So a sphere is used to show that the cube is expanding proportionally in all directions with respect to the X,Y,Z axes (these axes are all positive about the center). The axes are *aligned with the object* rather than the VSP because the VSP is symmetrical in all directions and orientation doesn't matter as far as expansion goes. The expansion axes for each VSP is aligned with the Z axis perpendicular to a plane that is tangent with the object's surface, and the X and Y axes are parallel with that tangential plane and orthogonal to each other. (I apologize for you nonscientific folks! Think of the VSP expansion sphere as a ball sitting on a flat and level table. The Z axis is a line that would go through the center of the ball and through the point on the ball that contacts the table. The X and Y axes would be lines that are parallel to the flat surface of the table and form a perfect cross where they intersect the the Z axis at the center of the ball.)

Fixed VSPs in open vacuum space, where they are not influenced by objects of matter, expand proportionally at the speed of light in every direction on average, as shown in Figure 6.3(a). It doesn't matter what the orientation of the cubical VSP is with respect to viewing orientation. Note that this diagram and the following three diagrams for VSP distortion are depictions of the *proportional expansion rate of the VSPs, not the shape of the VSPs*. The coordinate axes are graduated in Meters/Second as we see them from our perspective in the Universe. The VSPs are cubical in shape before distortion.

Proportional VSP Expansion Rate Diagram

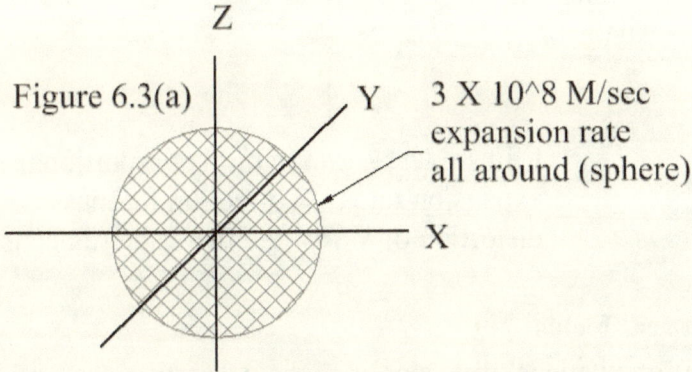

Figure 6.3(a)

3 X 10^8 M/sec expansion rate all around (sphere)

Proportional expanding rate of VSP is equal in all directions if not influenced by charged objects or objects with mass.

Gravitational Field

Gravitational distortion of the VSPs around an object is in the radial z direction, which would be perpendicular to the object's surface as shown in figure 6.3(b). Notice that this is a restrictive distortion so that the space pressure is reduced in that direction. This is consistent with our 2D model of the previous chapter so that gravity and inertia can still work radially as described in the simpler model.

Proportional
VSP Expansion Rate Distortion Diagram
Gravity & Inertia

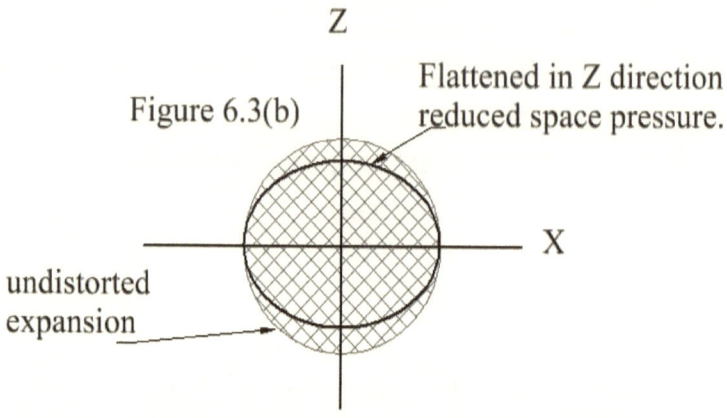

Figure 6.3(b)

Z

Flattened in Z direction
reduced space pressure.

X

undistorted
expansion

Ellipsoidal restricton of the proportional
expansion rate in Z direction for mass
distortion of VSPs (gravity & inertia cause).

Electrical Fields

If gravitational distortion is in the z direction, then the electric and magnetic VSP distortions would be in the planar x and y directions parallel to the object's surface. These distortions are probably not completely uniform around the objects and depend upon their position relative to the different groups of space particles comprising the charged object. But when added up, the resulting distortion would be predominately in the x or y plane that is parallel to the "surface" of the charged body. In a static situation (no movement) these distortions are passed from one VSP to another to form electric "field lines" around the object, diminishing as the radial distance increases.

The VSP distortions caused by a charged object are shown in figures 6.3(c) and 6.3(d). Notice that these distortions are *expansive* rather than restrictive, that is, the rate of expansion in the x or y direction is *increased* incrementally over the normal speed of light expansion in the VSP. (I want to mention here that the changes in expansion rate caused by charged objects that we normally would observe are very small compared to the normal speed of light expansion.) The two different and complementary expansions are caused by the two different positive and negatively charged objects. In both cases there is an incremental increase in volume of Universe space within the VSP as a result of the distortion, and this is equivalent to the differential energy that we normally see.

Proportional
VSP Expansion Rate Distortion Diagram
Positive Electrical Distortion

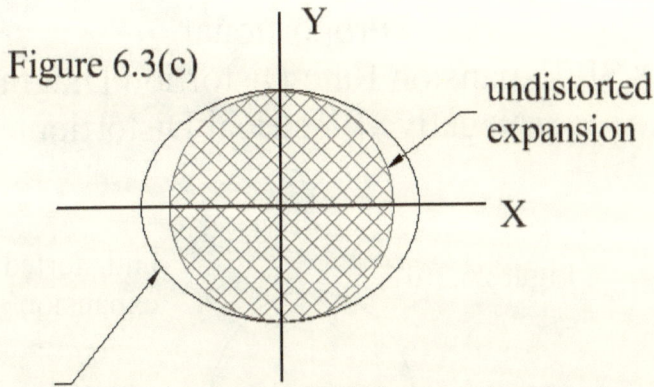

Figure 6.3(c)

Y

undistorted expansion

X

Ellipsoidal increase in X direction (proportional expansion rate for positive charge distortion of VSPs.)

In other words, energy is stored in the VSP electrical field around a charged object. We can't see the creative "absolute energy" that is causing the existence of the Universe, we only see the differential change in the rate. The electrical distortion, passed from one VSP to the next in a diminishing fashion, form the electrical flux lines that exist around the charged object. So, even though this flux is invisible, it is not magic and is still "real". It consists of distortions in the only "physical" component of the Universe!

Since the electrical distortion is expansive, it also *increases* the space pressure in the x or y direction. A charged object is surrounded by VSPs having a higher space pressure in the x or y dimension, depending on its positive or negative polarity. Like the gravitational z direction reduction in space pressure of VSPs surrounding an object, the x or y increase in VSP pressure surrounds a charged object with a net zero force on the object. Since the two electrical directions are orthogonal to each other and also to the gravitational z direction, the pressures only act on objects having differentials in VSP pressure in the same direction (i.e. x, y, or z). So now if two positively charged objects are brought into the vicinity of each other with the increased space pressure in the x direction, the addition of the two VSP pressures in the x direction on the side of the objects facing each other will apply an unbalancing force that will be in a direction to separate the two objects.

Proportional
VSP Expansion Rate Distortion Diagram
Negative Electrical Distortion

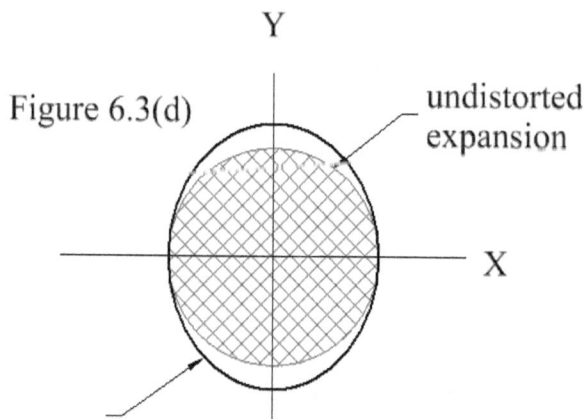

Figure 6.3(d)

Ellipsoidal increase in Y direction (proportional expansion rate for negative charge distortion of VSPs.)

Likewise, when two negatively charged objects having increased VSP space pressure in the y direction are brought in the vicinity of each of each, they also experience a force to separate them. And, if two oppositely charged objects are brought in the vicinity of each other, they will have forces applied from the VSPs on the sides away from the two objects which will push them together. This is due to the unbalanced lower pressures of the VSPs in the respective x and y directions on the objects sides facing each other.

Magnetic Fields

In the case of magnetic distortion, the distortions are produced in objects composed of atoms and molecules which have a much higher density of configured space particles, and these molecules have both compliments of the magnetic distortion so that they form dipoles, one side a north pole, and the other side a south pole. This magnetic distortion from the molecule is mainly caused by the relatively unbalanced movement of the charged electrons within it. And this in turn is caused by the arrangement and expansion of the clusters and sub-clusters of space particles within the nucleus of the atom. So the magnetic force is really produced by moving a charged body through the stationary VSP medium. It doesn't require that it be contained in an atom or molecule. An electron or proton can be "shot" through vacuum space independently as is the case in a vacuum tube, and a resulting magnetic "field" will exist encircling it in a plane orthogonal to the direction of travel. The requirement of moving the charged object through the VSP medium for the magnetic effect to exist suggests that the distortion of the VSPs as the charged object moves through causes a VSP x,y expansion "wake". This wake would be in a direction orthogonal to that caused by the charged object (like an oppositely charged object). And similar to inertia for massive objects, no net force will be applied to the moving charged object as long as the VSPs are allowed to "expand and contract" freely as the object moves through. If however another charged object is in the path of the electric "flux" and the VSPs interacting with it causes a restriction in the free expansion or contraction of the VSPs as the charged object moves through, then a force will be applied to the two objects. Also separately, the requirement of change in velocity of the charged particle to create the magnetic inertial force of inductance, and that the rate of change (acceleration of it) is proportional to the inductive inertial force, suggests that this force is something akin to the inertial force experienced for massive objects. In other words, the electrical force is proportional to the amount of charge on the two bodies and the rate of acceleration of the charged body. This relationship is expressed as Faraday's Law in classical Physics where the voltage induced in a conductor by a changing magnetic flux is opposite the flux inducing voltage and proportional to the rate of change.

The force on a charged object moving across the "field lines" (at right angles to them) of a magnetic field (x,y distorted VSP strings), is proportional to the velocity of the particle and in a direction that is at right angles to both the magnetic flux and the velocity path of the charged body. The same force could be applied to a stationary charged body by replacing the magnetic field with a static electric field whose strength was proportional to the velocity of the moving charged object and whose electric field orientation was orthogonal to the magnetic field in the x,y plane, and orthogonal to the charged object's path.

We know from classical physics that the direction of the magnetic field encircling a moving charged body with respect to its direction of movement depends upon the polarity of the charged body (positive or negative). We know that moving a charged body in a direction orthogonal to a magnetic field will cause a force to be applied to it. We know that moving a wire containing "free" electrons in a direction orthogonal to the magnetic "flux lines" will apply a force to the electrons and cause them to move in the wire. If the electrons are then allowed to flow in a circuit, a force resulting from the restriction of free expansion of VSPs will act on the magnet and the wire (this is how an electrical generator works). All of these actions are no more than just the forces of electrical attraction and repulsion on the objects caused by distortion of the VSPs in the x or y direction. So in reality there is just the VSP electrical x,y distortions acting on the charged objects and no separate magnetic force as such.

In summary, I have coarsely shown how the gravitational and electromagnetic forces applied to objects of matter could be the result of expansion distortion in the VSPs surrounding them and the resulting space pressure imbalance. Electrostatic, magnetic, and gravitational "fields" , and "flux", would consist of strings of distorted VSPs which are passed one to the other around the object in a diminishing fashion. The VSP distortions are passed one to the other by virtue of the restricting boundary conditions between them, and they would be the sole cause of the invisible mysterious forces acting on objects of matter at a distance. What profound thoughts! This would be cause for Isaac Newton to be dancing a little jig with Albert Einstein about now if it is true, but a great deal of work must be done to confirm that this is really so.

Won't you join the party?

Einstein and Newton Discover Bubble Ether!

And now we enter into this last chapter, the lucky seventh. Here we will glimpse a little deeper into the quantitative chaotic expansion of space particles, resultant time, and energy as we see it.

Remember that matter consists of groups and subgroups of compacted yet still expanding space particles which form the familiar subatomic particles of matter that we can observe. Filling the void between these subatomic particles is the stationary and invisible inner space VSPs with their x,y,z distortions that bind the subatomic particles together through boundary conditions and vacuum space pressure. We are able to detect the presence of the larger groups of space particles like Protons, Neutrons, and Electrons with instruments we have made. We see that they have mass, a physical size, and hold a unit charge, or are neutral. These groups behave according to the laws of classic physics as long as the groups are not accelerated to any significant portion of the speed of light in the stationary VSP medium. These atoms and larger components of such, are just stable configurations of space particles in a stable state of expansion that is in proportion to the VSP expansion in our region of the Universe. But remember that the smaller clusters of space particles in the atom (subatomic particles like quarks) may be expanding "herky jerky" at different rates within the nucleus, and expanding in discrete steps that are random and chaotic with each other. So what are we really observing when we "see" an electron or proton with a constant size? We are observing a group of space particles whose *average* expansion rate ratio over the observation period is constant with the *average* expansion rate of the reference VSPs in our region.

Energy

The sole creative force driving the existence of the Universe is the expansion of space through the creative expansion within each of the individual space particles. As long as it exists, the total volume of the Universe is ever increasing in the infinite Void, and its current value represents the total absolute energy of the Universe at any given point in its expansion.

Assuming that the infant Universe was spherical and is composed of a finite and constant number of space particles, and its initial diameter is the constant reference of length used in the Void (Princh unit), then the initial volume of the Universe, **Vol_I**, would have been the volume of a sphere with a one princh diameter. This can be found from Equation 7.1:

Equation 7.1 $$Vol_I = \frac{\pi}{6} \, prinches^3$$

The total absolute energy in the current Universe then could be defined as the ratio of its current volume to its initial volume in absolute energy units (prinches³). Now the number of space particles is finite and has remained constant during this expansion, so a ratio of volumes is a ratio of the expansion of space. From this we can say that any given volume of Universe vacuum space (finite and fixed quantity of space particles) contains absolute energy that is equal to a ratio of its current absolute volume in prinches³ to the initial volume Vol_I. It follows then that the amount of absolute energy in a given current volume of Universe vacuum space would be equal to its current absolute volume in cubic prinches divided by π/6. Of course we cannot see this absolute volume, or measure it directly. The expansion rate is most likely not uniform over the whole Universe either, so it may be hard to glean the rate from the speed of light indirectly as well.

What we see as the exhibition of energy in our visible Universe is caused by an *incremental change* in the expansion rate of an observed volume under consideration relative to the expansion rate of a reference volume of vacuum space. If we represent the "normal" vacuum space expansion with ΔV_N and the energetic volume expansion change by $\Delta(\Delta V_E)$, where **K** is a constant of proportionality, then we could define the magnitude of our visible energy exchange resulting from the increment of expansion E_V, by the formula:

Equation 7.2 $$E_V = \frac{(\Delta(\Delta V_E))}{(\Delta V_N)} K$$

If the ratio is positive, then energy is released from that volume. If it is negative, then the energy is stored in that volume.

We normally cannot observe absolute energy, because most of the invisible one way volume expansion is "used up" causing the physical properties of the Universe and our physical existence (Could you see the inside of your only eye with that eye if it didn't function while you were viewing inside it?). What we do see as energy and the expansion of space is a differential (rate of change) of expansion relative to the background VSP expansion. This relationship is constant, and so it appears to us that what we have assumed to be the energy of the Big Bang is always conserved.

We have talked about matter and the complex configurations of groups and subgroups of space particles in them (which can be most any polyhedral shape vs. the cubical VSPs). The configured space particles within any object, charged or uncharged, expand in their x, y, and z dimensions in quantum steps, and on average expand inversely proportional to the density of the object they construct. If there is an incremental increase in this overall expansion rate relative to the "normal" VSP expansion rate, then the relative increase in volume that results with expansion relates directly to the amount of energy displayed. When the nucleus of an atom is "smashed" by the collision with a high velocity proton or electron, the stable expansions of its subatomic particles are disrupted and the x, y, z configuration bonds are broken, allowing the space particles in them to expand at a higher rate than they were in the original configuration. This burst of expansion causes a volume increase in the space particles above the configured volume (like an explosion), launches "rays" of very short and intense electromagnetic radiation (gamma rays and such) into the VSP medium, and ejects subatomic particle shrapnel. The subatomic particles ejected may collide with other nuclei and excursions in expansion may be so intense in the VSP medium that they themselves may cause destabilization of subatomic particles in other atom nuclei in the vicinity, resulting in secondary radiation and deterioration of the nucleus into a different elemental configuration. Some of the ejected subatomic particles with properties of mass become so destabilized in the collision that they break into freely expanding VSPs, loose their mass as a result, and disappear from our view entirely even though they still exists as VSPs (mass is converted to energy).

For electromagnetic energy that is "traveling" through vacuum space, the increased expansion rate in the affected VSPs is simultaneous in both the x and y directions. This expansion results in an incremental change in the VSP volume relative to the steady VSP expansion volume, and manifests itself as electromagnetic energy. This energy distortion ripple "travels" through vacuum space, passed particle to particle at the VSP expansion speed which we see as the speed of light. The static electric and magnetic distortions of the VSPs separately do not result in the launch of an electromagnetic wave that can travel independently through vacuum space. Instead the incremental volume change is stored as energy in the "flux" (VSP distortion strings) surrounding the object that produced it. It is only when the x and y expansion distortions are simultaneous and changing at some period rate that a wave is launched. The rate of change relates to the frequency of the electromagnetic energy, the magnitude of the excursion relates to the intensity of the energy burst. All electromagnetic energy emanating from an atom, whether it be infrared thermal energy up to x rays and higher, result from a change in the atom's nucleus in which the configured space particle groups and subgroups shift to cause an incremental change in expansion above the regular expansion.

Heat Energy

We have mostly been talking about energy from an expansion of space particle perspective, which is ultimately the origin of it, but more often we see the resulting incremental expansion of atoms and molecules exhibited as thermal energy. The chemical reaction type of incremental expansion rate that we normally encounter and can see when fuel is burned is very much lower in magnitude than the VSP expansion rate or that when a clump of space particles (matter with mass) is converted into VSPs by an increment of expansion rate to that of the VSPs ($E = m \cdot c^2$ nuclear explosion type of energy).

When you burn a fossil fuel, oxygen molecules are combined with hydrocarbon molecules in a chemical process with "heat" that allows the molecules to exchange elemental atoms and form new molecules and compounds such as carbon dioxide and water. These compounds are just a different configuration of weakly bound elemental atoms that have been formed as the agitation of the involved atoms subsided or "cooled". As this happened "heat" was exchanged with the surrounding atoms and molecules causing the process to sustain itself as "burning".

But what is heat? What does it mean to say that something is hot? Basically it is the agitation of the electrons, atoms, and molecules, and by this I mean how much, how fast, and how far do they move around in the VSP medium over a period of time. Much of what we see in heated gases and solids is kinetic energy (the energy of motion). Remember when we were talking about inertia back in chapter 5?

When an object is moving in the VSP medium, the energy that it took to get it moving is stored in the "compressed" VSPs in front of the moving object, and it is liberated whenever the motion is slowed down and the VSPs are decompressed (work is expending energy and is equal to applying a force through a distance). In a gas such as hydrogen, the hydrogen atoms are not bound together in a rigid configuration like they are in a solid, and they are free to move around. The temperature of the gas indicates how much the molecules are moving around, how fast they are moving, and the level of the kinetic energy of motion in them. The molecules impinging upon the impenetrable walls of an enclosure is what creates the force and pressure of containing the gas. (Now I will mention here that the kinetic energy of an atom or molecule in a gas or solid at a given temperature varies randomly over a range of energy levels, and when we talk about the kinetic energy of a gas we mean the statistical average of kinetic energy in the molecules at a given temperature).

In a solid, it is the cloud of electrons (like a swarm of bees) surrounding the nucleus that gain velocity, "jump" to higher "shells", and take up a bigger volume as the temperature increases. This causes the physical expansion of the solid as it is heated. There may also be some shifting of position or even some cyclic shifting going on inside the nuclei of the molecular atoms as the temperature rises as well. The kinetic energy, or heat energy of the solid, is stored in the motion of the electrons about the nuclei of the atoms.

There is also another state of matter called a plasma in which the nuclei and electrons of the atoms become a "soup" at extreme temperatures. In this state, the kinetic energy is so high that the electrons break the bonds they have with their nucleus and "swap" places with each other. In this condition, the electrons and positive ions (nuclei stripped of electrons) can move in an electric field (distorted VSPs that were mentioned in chapter 6). This is what happens in a welder's arc. Kinetic energy is gained in the heavier nuclei as they are accelerated in the electric field, and this is transferred to the cooler metal as heat when they impact it. The energy supplied to produce the heat came from the welder electric current.

When molecules with the same kinetic energy or temperature collide with each other, the collisions are "elastic", that is, on average no kinetic energy is transferred between them. If the molecules are of a different temperature, then kinetic energy is exchanged between them until an equilibrium is reached. The transfer of kinetic energy action is called heat flow, and naturally moves from a hot object to a colder one.

Entropy

In order to extract thermal energy from objects to do useful work for us, there has to be heat flow. The natural flow is always from a hotter object to a cooler one (stated as fact since the opposite of this has never been observed). In practice, both objects have heat energy as indicated by their *absolute temperature* and capacity to hold heat (*specific heat* of the object). If two objects of different temperature are brought together physically in an isolated environment where none of the heat energy from either object can escape, then heat will flow from the hotter object to the cooler one until the temperature of the two objects are equal. This natural flow is irreversible, that is, heat cannot be made to flow from a colder object to a hotter one without supplying external energy to the system. The total energy of the closed system is now distributed between the two objects at the new equilibrium temperature, and the heat flow ceases. When this happens, no useful energy can be extracted from the heat energy contained in the two objects, even though they still contain the total energy that was present before they were brought together. Useful energy can only be extracted during the process of heat flowing from a hotter object to a colder one. This is because only the difference between the kinetic energy (energy of movement) in the molecules or electrons in the hotter object and the cooler one can be converted to another form of energy during the transfer. Also, if some other form of energy is used to heat an object, the total energy used to heat it cannot be extracted back from the object by any process without using external energy. This phenomenon is essentially the Second Law of Thermodynamics, and its implications are that the efficiency of any device or process to extract energy from the environment to do any useful work will always be less than 100%. This lost, unrecoverable energy phenomenon is also commonly called entropy. This natural behavior, much to the chagrin of perpetual motion and "free energy" machine hopefuls, is the demise of their cleverly designed devices. It is true that all things physical have energy flowing into to them from a source outside of our Universe as my Metaphysics and New Age friends have proposed. And, their hearts are in a good place to want to tap into this for the benefit all people if we could use it. But unfortunately, this energy is the invisible absolute energy, which is the one way expansion of space volume, and it is "used up" causing us to exist in the "now". We cannot tap directly into that energy for other uses.

All objects with a temperature above *absolute zero* (point where no kinetic energy exists in the object) loose some of their energy as radiant heat (a form of electromagnetic energy). It may be that they are receiving as much radiant heat energy from other atoms surrounding them so that their temperature remains constant, but they are radiating just the same. On the surfaces of a hot object in an empty vacuum, the radiant heat energy leaves the object and is lost. There are no other objects around with a higher temperature to receive radiant heat from, and this results in cooling of the object. The electromagnetic wave front spreads out as it leaves the object and the radiant energy disperses over a wider area. So if this wave encounters another object of the same size at a distance, then only a portion of the original amount of radiant energy is gained by it. The rest is lost into Universe space as an electromagnetic wave and results in vacuum space being filled with electromagnetic waves of all sorts!

The temperature or kinetic energy within an atom may be changed in two ways. One way is mechanical in which the kinetic energy of the atom or molecule is increased or decreased by contact with another of a different energy level as we have just considered. This method doesn't require any action from the VSPs other than that causing the storage of energy in motion, and is well understood in classical physics (Thermodynamics). The other method of changing the kinetic energy and temperature of an atom or molecule is through electromagnetic radiation. In this method a portion of the molecule's kinetic energy is transformed into a change in the incremental x-y expansion rate of the VSPs surrounding it. Cyclic movements within the atom launch these x-y expansion ripples in the VSPs which are then passed from one to the other at the speed of expansion (speed of light) as an electromagnetic wave. As the wave was launched in the originating atom, its temperature cooled due to the loss of some of its kinetic energy. Also, if an electromagnetic wave encounters a remote atom, the expansion ripple in the VSPs will set up a cyclic movement in it, causing its kinetic energy and temperature to rise if its radiant energy is less than the amount it receives. This is how we receive all of the heat energy from the sun.

Time

I mentioned back in Chapter 3 that time passes as a result of the linear expansion of space. Now we can say that time passes as a result of the independent volume expansion of absolute space, and is a scalar quantity (has no special direction to it) that can be expressed functionally as the linear expansion of absolute space in terms of absolute volumetric expansion. Okay, that was a mouthful (*sorry*), so what does that mean to us ordinary folks? Well, back in the previous section on energy, we talked about the absolute energy in a given spherical volume of absolute space, and how an increase in this volume due to expansion equated to a direct increase in absolute energy. Now we are going to see how an expansion in the volume of absolute space will translate into a linear expansion of it. The result will give us time as a function of absolute space expansion. Now for those of you with Math-phobia, don't "freak out" and get discouraged by the following little exercise on describing time mathematically. It is sufficient to know that time happens as a result of the creative expansion of absolute space. Just tip-toe over this and remember that you can describe time in terms of space expansion if necessary.

Again we will work with a spherical volume of space for ease of understanding and calculating since the shape of the volume under consideration is arbitrary. In this case the diameter of the sphere will be used as the linear measurement subject, and its increase with the volume increase of the sphere is the sought after item. The volume of a sphere V_s in terms of its diameter $\mathbf{D}$ is:

Equation 7.3 $$V_s = \frac{\pi}{6} \mathbf{D}^3$$

Solving this for $\mathbf{D}$ as a function of V_s we obtain:

Equation 7.4 $$\mathbf{D} = \sqrt[3]{\left(\frac{6}{\pi}\right)} \cdot \sqrt[3]{(V_s)}$$

So, from this we can see that if time passes as a result of the linear expansion of space, which would also be directly related to the change in the diameter of the spherical volume being considered, the amount of volume expansion required to make each second pass is ever increasing with each increment of expansion. This increasing amount of absolute energy required for time passage seems odd, but it must be so if time really is the result of absolute linear expansion in a volume of Universe space.

So, if we wanted to see how much time passed in a volume of Universe space during an interval of absolute volume expansion, we would have to know the expansion interval's starting volume V_{A1} and its ending volume V_{A2} in absolute volume units (prinches³) Then the time that has past, **t,** would be directly related to the change in the diameter of the spherical volume **ΔD** multiplied by some factor **K** to get it into seconds. Equation 7.5 expresses this in mathematical form:

$$\mathbf{t = \Delta D * K = K * \sqrt[3]{(\frac{6}{\pi})} * [\sqrt[3]{(V_{A2})} - \sqrt[3]{(V_{A1})}]}$$

Equation 7.5

If we knew the total absolute volume of the Universe at any given instant, we could calculate the average time the Universe has been in existence by using its initial volume of $6/\pi$ as V_{A1}.   But from our existence point in the Universe, we can only deduce what the dimensions might be from our observations. There is just one connection between the Void absolute Princh units and our expanding Universe meter units, and that is the rate of the average expansion of vacuum space; the speed of light.   But this might not be consistent throughout the Universe ... we may never know.

Uncertainty in minutia

To observe anything, expansion must occur and resulting time must pass. Indeed, expansion must occur for anything of the Universe to exist! (A profound statement to contemplate!) In order to observe an action, we must view a volume of our expanding Universe over a period of its expansion. The expansion causing time to pass for an object being observed is an average expansion of all the VSPs within it, (i.e. the diameter in Equation 7.4 involves many clumps and clusters of VSPs in the object).   And this is referenced to the average VSP expansion in the volume of our measurement standard which we consider constant. Since the expansion of small clumps of matter, such as subatomic particles, is the result of randomly occurring expansion of its space particle components, then it is logical to conclude that the position of a subatomic particle might be uncertain at any given instant in time. In fact, time as we see it has no meaning in the subatomic particle sense, because we are viewing it from an average expansion of a much larger volume of space. This averaging smooths out all of the little quantum steps in volume change, and we do not have the resolution to see these smaller increments of expansion. This causes the uncertainty of where the little particles might be when we try to view them in a very short period of time.

Werner Heisenberg

Das Bean ist Ungewiss!

(The Bean is Uncertain)

Epilogue

Recapitulation

- The entire Universe consists solely of space particles. Each particle is unique in that it cannot share its space material with any other particle. These particles might be thought of as Higgs bosons, except that each particle contains a creative source of its unique space material that is forcefully trying to expand the particle. Where the material comes from is unknown, and it is uncertain how long it will continue to supply new space material.

- There are no voids between particles, so the particles must share sides at all times. Singular particles are invisible to us, and are called vacuum space particles (VSPs). They are the fastest expanding particles, and expand at the speed of light. This expansion rate might not be constant throughout the Universe. A volume of VSPs could be considered an expanding Higgs field.

- Clumps of space particles can form, and these expand at a slower rate than VSPs. These clumps have properties of mass and are visible in the Universe. The mass is proportional to the density of the clump (relative to a VSP).

- All actions of the Universe are the result of the expansion of invisible space particles. Our awareness of the Universe's existence comes solely from the visibility of these actions even though we cannot see the space particles or observe them expand.

- Time passing is a record of linear expansion of space particles in absolute length units (non-expanding units seen in the Void). The space particles can only expand, never contract or completely stop expanding, and therefore time only moves forward or can only approach a stop.

- Universe distance measurement is no more than a comparison of a quantity of expanding reference space particles to a quantity of expanding space particles in the objects that are being measured, or that are displaced in vacuum space.

- Invisible Absolute Energy is the volume expansion of space particles into the Void in absolute length units cubed, and it is continually being created. We cannot see this energy.

- The visible exchange of energy that we observe is the result of an incremental expansion rate change in a volume of space particles being considered to the expansion rate of a reference volume.

- In the case of electromagnetic energy, the incremental volume of expansion of the originating object ripples through the VSP medium at their expansion rate (which is the local speed of light).

- There was no "Big Bang" in which all the energy of the Universe was created at its beginning. Energy is not "conserved" in the sense now thought with that theory. What appears to us as conservation of energy results from the constant relationship and connection between time (or linear expansion), space (quantity of VSPs) and change in volume expansion rates (Universe observable energy).

- All forces of action on objects at a distance, such as gravity, inertia, or electromagnetic forces, come from an imbalance in the vacuum space pressure. The space pressure might not be constant throughout the Universe, and may vary according to location and absolute distance from the Void boundary.

- If the source material within each particle comes from the same place outside of our Universe, and there is a finite quantity of it (likely to be true since it had a beginning), then at some point all expansion will stop. If this happens, nothing of the Universe will exist, and there will be no record that it ever existed. Time as we know it is only a property of the Universe and will no longer exist.

Conclusion

I have always had an intense desire to know how things work. I have been blessed (or cursed) with a child-like curiosity and imagination that has driven me to spend my life trying to understand how and why things behave the way they do. As an engineer, I have used my knowledge and imagination to invent devices and use Mother Nature's behavior in ways that are beneficial to us. Also, growing up in a rural environment where we had a close connection to Mother Earth and depended upon living plants and animals for our food and livelihood, I have developed a deep respect for life, and I have seen and felt the spirit and soul in living things. Words fail me here, but the closest thing I have to describing this is love. I have come to believe that all living things share and reflect a portion of God's love (yes, I believe in God), and that this is the essence of our consciousness, soul, and spirit.

Some years ago, I woke up from a dream in the middle of the night; a vision in which I saw the Universe as an expanding foam. I got up and wrote these thoughts down in the form of a message from a metaphorical Sparrow. With my limited knowledge of Physics and Mathematics, I have tried to understand how this foam could cause the existence of our Universe. The more I work on it, the more sense it makes, and the more I believe this is the actual truth. In my dream vision and subsequent contemplation of Universe workings, I've gone into the real world where there is just the spirit of all things. The facade of solid objects, the physics of their actions, and time, is just an illusion. It is all made up of invisible space bubbles! The only thing making it all exist is the breath of existence that is continually being given.

I have tried to make this believable for you with some scientific explanations of how it could work. I might have made a few technical errors in doing so, but my work is not to "prove" all of this. When I received the vision, I was given the order to be a messenger and spread the news to those who would listen. I do not seek fame or fortune ... I'm an old man, and my life is about over. My mission is to present my vision to other dreamers and believers so that they can share in the exciting work of seeing how it all happens. I hope that writing and publishing this tiny little book about our great big Universe has done that. And hopefully from this inspiration and through their work and efforts, we will be able to raise mankind's understanding of our existence to a higher level where we can live in peace and coexist on this Earth with the rest of creation.

If it is true that the Universe is no more than a set of expanding space bubbles as I have proposed, with some work and deductive reasoning, we will be able find out the cause of all physical actions of the Universe. We will see that our Universe is in a constant state of creation, and all things associated with the physical is caused by and dependent upon the creative growth of new space material within each of the space bubbles. If the space material filling the bubbles "runs out", the time for time will have run out. There will be no matter, energy, gravity or inertia, or even time; no evidence that the Universe ever existed. But even with this knowledge, we will not know where the continuing creation of unique space material within each of the space particles comes from, or why the particles exist in the first place. We still will not know how life began and what causes it to evolve in a physical Universe that is not affected by the spiritual. We will not know how our consciousness developed, or from Whom it all came.

And so now we have come to the end of my book. My intentions here have been to open your eyes and arouse your interest to the possibility that the Universe could be existing in a completely different way than we have thought up until now. I have used simple examples to illustrate how some of the heretofore mystery forces of gravity, inertia, and electromagnetism can act on bodies of matter at a distance with invisible expanding space particles. I have shown what time, space, and energy could be, and have shown how they are interconnected. I have done so without any proof that what I propose is actually true, other than it is logical to think that these things could be caused by expanding space particles if they really exist.

To those of you with an open mind, these revelations should ignite an intense fire in you to explore the possibilities of completely understanding how Mother Nature works. It is like standing beside the ancient Philosophers and Mathematicians as they wrote their philosophies, like floating with Archimedes when he shouted "eureka", like peering through Galileo's telescope or observing Kepler's sightings, like being there with Isaac Newton when he saw an order and predictability to movement and gravity. The work done previously recording the behavior of Mother Nature is still valid, for it is as we see it in our Universe! It is just now we can see that *it is happening for a different reason than we thought.* If the expanding space particles are really the cause of the Universe's existence, then there must be a way to show how they do what they do.

Listen and hear you brilliant minded Physicists, Mathematicians, Cosmologists, and Engineers who are bored out of your minds, and who like some before you think that "everything there is to know about science is practically known", and who think that there's nothing left for you to exercise your mind on; here is some meaningful and exciting work for you! Humbly take the knowledge that those before you have gained, open your eyes wider now with this new vision, and use it to come to a more complete understanding of what it is that we are physically, and how we exist in the Universe. But remember that we might never know how it was that our consciousness, spirit, and soul got here, or what is to become of us. *Maybe God is blowing the bubbles that makes everything exist!*

And for all of you Prophets, Spiritual Philosophers, Teachers, and Reflectors of the Creator's love, if you are humble and courageous enough to peek behind your religious veil, you might discover that you were right all along to think that God is love and that God is in all of creation. It's just that you were near-sighted and could only see a small part of the truth to think that we are the center of the Universe in a spiritual sense. For if God is breathing life into our Universe for every second that passes, and is the essence and source of our consciousness and soul, we need to be humble and grateful for our existence! It is our arrogance and narrow mindedness that creates Hell here on our tiny Earth. We will have to humbly admit that we are not here by our own doing, and that our knowledge of the workings of the Universe has no bearing on its existence. And when we open our eyes to this, we will see that the Scientific, Religious, and Metaphysical Philosophers have been like the Three Blind Mice trying to describe an elephant based solely on the little part they are in touch with. We are a part of this glorious creation, not the cause, and we must peacefully coexist with all of it. We will just have to trust and have faith that our Creator will guide us into the light.

Sparrow Whispers

Just this morning 09/12/2010 at 1:30am, a metaphorical sparrow awoke me with a message. She said that she was sent by the One who knows, and that I must get up and write this down accurately, lest I forget or discard this as a silly dream, and that I must spread this as a second messenger to all who will listen. Here is what the sparrow said:

I am sending this message through the simplest of means so that all who seek the truth with humility can see it, and can know the answers to questions that plague their minds. I do this because arrogance has blinded all that possess it, and has been the original cause of all your discord.

What are you?

You are only two things of all possibilities. One portion of you is physical and is bound to the first thing, which is the Universe. The other portion is your intangible spirit, which is a second thing, and is similar to the laws that govern the Universe, or in modern terms, software. Your two parts are separate and cannot affect each other directly, even though they both are embodied in you.

Your Universe is just one finite thing of many identical but individual pieces, and all of its apparent properties; time, matter with its own properties, space, and energy, come from one unifying source on a continuing basis. Its existence is the result of what you think of as the expansion of space. It had no beginning in the time sense, because time is only a property of the Universe.

Your spiritual intangible part is your conscious mind that is aware, can reason and understand what I am saying. Like the Universe, your spirit is continuing to be created and your knowledge and awareness grows. You can use spiritual tools like mathematics, to better understand the behavior of the Universe, but your understanding and awareness of those laws still cannot affect them in any way, nor can be the cause of any existence. Your confusion in this matter comes because the behavior of the Universe is governed by a spiritual type law which comes from outside of it, and is similar to your own spiritual nature.

As for why you exist ... because I love.

It is now 3:00am, and I am finished writing. Going back to bed. Will attempt to spread the message in the morning.

Octahedron as a Component of a Solid Analysis

First:

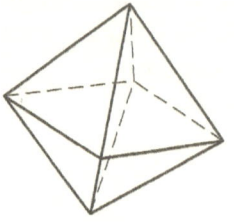

$$h^2 = S^2 - (D/2)^2$$

$$D = S*SQRT(2)$$

$$h = S/SQRT(2)$$

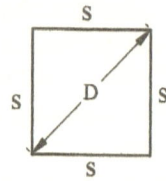

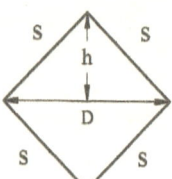

Octahedron

Section thru four edges & four apexes to get diagonal D

Obtain height h

Then, cross section the hexahedron through two apexes and bisecting four of the sides to find the angle made when the faces of the octahedrons are placed together to form a ring.

7.5 deg

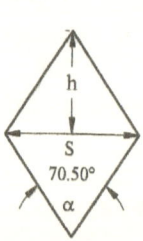

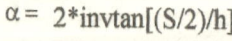

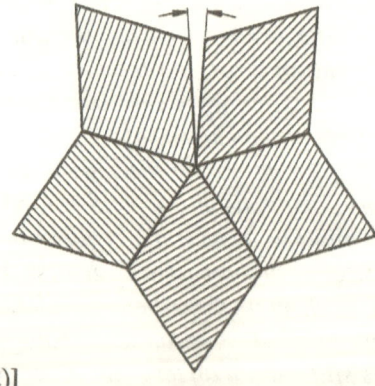

$\alpha =$ 2*invtan[(S/2)/h]

$=$ 2*invtan[(S/2)/(S/SQRT(2)]

$=$ 2*invtan{[SQRT(2)]/2}

When the hexahedron facets are placed face to face to form a solid ring, a 7.5 degree gap exists in the ring between two of the faces. Therefore the hexahedron cannot be the shape of the vacuum space particles (VSPs) because no gaps can exist between them.

Tetrahedron as a Component of a Solid Analysis

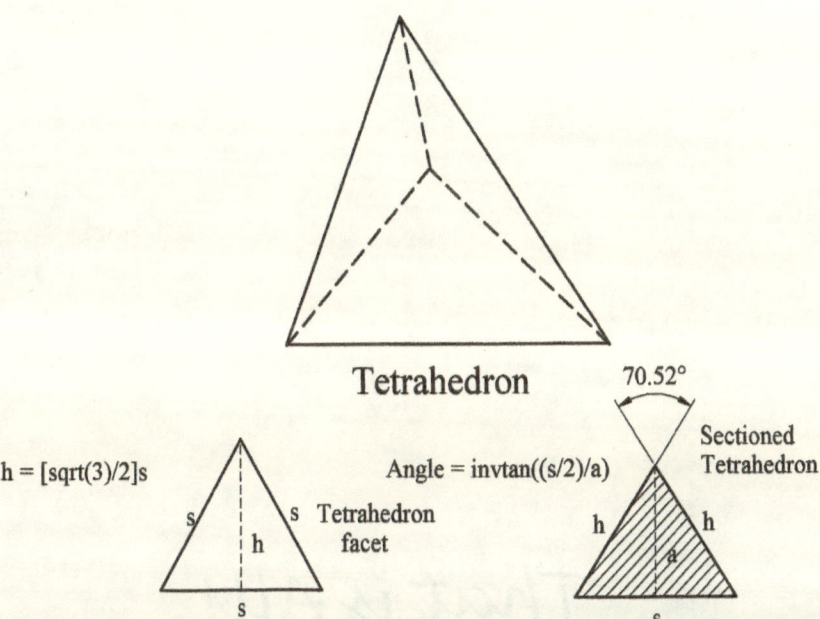

Tetrahedron

$h = [sqrt(3)/2]s$

s s Tetrahedron facet

h

Angle = invtan((s/2)/a)

70.52°

Sectioned Tetrahedron

h h

a

s

Find height (h) of a tetrahedron facet by the Pythagorean theorem and then section the tetrahedron through one edge and along the height bisecting line of the two opposing facets so that the height lines form the two sides of the section isoscoles triangle. The angle formed at the apex of the isoscoles triangle must be an exact submultiple of 360 degrees if the polyhedrons are to fit together perfectly without gaps or overlaps.

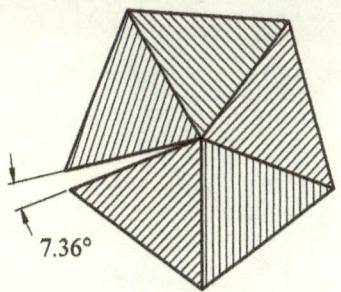

7.36°

When the tetrahedrons are placed together, facets face to face to form a ring, the cross sectioned pieces along the center plane of the ring must fit together without any gaps or overlaps. In this case there is a gap, so the tetrahedron cannot be the shape of the vacuum space particles.

That is All!

Let us Begin!